SPALLATION NUCLEAR REACTIONS
AND THEIR APPLICATIONS

ASTROPHYSICS AND
SPACE SCIENCE LIBRARY

A SERIES OF BOOKS ON THE RECENT DEVELOPMENTS

OF SPACE SCIENCE AND OF GENERAL GEOPHYSICS AND ASTROPHYSICS

PUBLISHED IN CONNECTION WITH THE JOURNAL

SPACE SCIENCE REVIEWS

VOLUME 59

SPALLATION
NUCLEAR REACTIONS AND
THEIR APPLICATIONS

Edited by

B. S. P. SHEN AND M. MERKER

University of Pennsylvania

Supported in part by the National Aeronautics and Space Administration
and the National Science Foundation

D. REIDEL PUBLISHING COMPANY

DORDRECHT-HOLLAND / BOSTON-U.S.A.

Library of Congress Cataloging in Publication Data
Main entry under title:

Spallation nuclear reactions and their applications.

(Astrophysics and space science library; v. 59)
Expanded and updated versions of papers at a conference held
May 1975 at the University of Pennsylvania.
Includes bibliographical references and index.
1. Spallation (Nuclear physics)—Congresses. 2. Astrophysics—
Congresses. 3. Cosmic rays—Congresses. I. Shen, Benjamin Shih
Ping, 1931– II. Merker, Milton. III. Series.
QC794.8.S7S67   523.01′9′754   76–21803
ISBN-13: 978-94-010-1513-4   e-ISBN-13: 978-94-010-1511-0
DOI: 10.1007/ 978-94-010-1511-0

Published by D. Reidel Publishing Company,
P.O. Box 17, Dordrecht, Holland

Sold and distributed in the U.S.A., Canada and Mexico
by D. Reidel Publishing Company, Inc.
Lincoln Building, 160 Old Derby Street, Hingham,
Mass. 02043, U.S.A.

TABLE OF CONTENTS

Nine years ago, in 1967, a conference on spallation nuclear reactions and their applications in astrophysics was held at the University of Pennsylvania.[1] Since that time, a number of developments have given renewed impetus to the study of spallation reactions. Among these are the successful acceleration of high-energy heavy ions in the laboratory and their potential use in cancer radiotherapy, the availability of returned lunar rocks containing records of past cosmic-ray irradiation, and the development of the theory that the spallation of interstellar matter is responsible for much of the observed universal abundances of the rare light nuclides. In May 1975, a new conference on spallation nuclear reactions and their applications to astrophysics and radiotherapy was organized and held, again at the University of Pennsylvania. The papers in this volume are primarily expanded and updated versions of invited papers given at that conference.

To the authors of the papers, we owe a debt of gratitude for their contributions and for their forbearance. The conference itself was much stimulated by the services of the four session chairmen: William A. Fowler, Serge A. Korff, Robert Serber, and Maurice M. Shapiro, each of whom has over the years made fundamental contributions to the subject matter of this volume.

Crucial support for much of the editorial work was provided by the National Aeronautics and Space Administration and the National Science Foundation. George W. Pepper, Associate Treasurer of the University of Pennsylvania, was particularly helpful in expediting the final technical arrangements for publications.

Philadelphia, Pennsylvania
February 1976

B. S. P. Shen

M. Merker

---

1. High-Energy Nuclear Reactions in Astrophysics, edited by B. S. P. Shen (New York: W. A. Benjamin, Inc. 1967)

B. S. P. Shen is the Reese W. Flower Professor of astronomy
and astrophysics, chairman of the Department, and director
of the Observatory at the University of Pennsylvania.
M. Merker is Assistant Professor of astronomy and astrophysics
at the University of Pennsylvania.

B. S. P. Shen

Astrophysics Laboratory and Department of Astronomy,
University of Pennsylvania, Philadelphia, PA   19174

1.   SPALLATION NUCLEAR REACTIONS

This volume of collected articles deals with the type of
nuclear reactions commonly called "spallation reactions" and with
their occurrence in astrophysics, geophysics, radiotherapy, and
radiobiology. There is no generally accepted definition of the
term "spallation reaction"; usage differs somewhat from discipline to discipline. Astrophysicists, who have little interest in the mechanism of the reaction but great interest in
its outcome, tend to use the term with more abandon than the
nuclear chemists, who over the years have done most of the systematic research on the process at high-energy accelerators. Cosmic-
ray physicists, who have dealt with spallation reactions for
decades, continue to refer to such reactions suffered by cosmic
rays as "fragmentations", to the dismay of nuclear chemists who
reserve that term for a very specific and poorly understood
aspect of spallation. But this lack of uniformity in terminology
need not worry us here. For all practical purposes, spallation
reactions are inelastic nuclear reactions in which at least one
of the two collision partners is a complex nucleus and in which
the energy available well exceeds the interaction energy between nucleons in the nucleus. Thus, a nucleon-nucleus or pion-
nucleus or nucleus-nucleus reaction in which the incident energy
exceeds something like 50 or 100 MeV per a.m.u. is generally
referred to as a spallation reaction. The term comes from the
verb "to spall", meaning to chip with a hammer, but actual usage
has never required that there be wholesale chipping away of
nucleons before a reaction can be called a spallation. There is
no clear line separating spallation reactions from the lower-
energy nuclear reactions; rather, one type gradually merges into

Shen/Merker (eds.), Spallation Nuclear Reactions and Their Applications, 1–8. All Rights Reserved.
Copyright © 1976 by D. Reidel Publishing Company, Dordrecht-Holland.

the other as the incident energy increases and as the compound-nucleus model, so successful at lower energies, gradually fails to describe the process. In this introductory chapter intended mainly for scientists who are new to spallation reactions, I will try to set the scene for the various specialized chapters to follow and smooth over the discontinuities between chapters as they occur.

Spallation reactions are today described in terms of a two-step model originally due to Serber.[1] In the first step, the incident high-energy particle — say a 500 MeV proton — enters the target — say a Si^{28} nucleus — and interacts with some of the individual nucleons in the target in what is known as an intranuclear cascade. In so doing, a few nucleons will be knocked out of the target nucleus. If the incident energy is high enough ($\gtrsim$ 300 MeV), pions will be created and some of these too may be emitted along with the nucleons. The residual target nucleus, now containing fewer nucleons but raised to an excited state, then de-excites in the second step of this two-step model. The exact nature of the second step is not well understood, but the prevailing view holds that the residual nucleus de-excites by emitting, through the usual evaporation process familiar in low-energy reactions, a number of single nucleons and small clusters of nucleons, leaving behind a final nucleus. To give an example, the spallation reaction induced by our 500 MeV proton incident on the Si^{28} nucleus might lead to the emission, over the two steps, of a total of, say, 4 single neutrons, 3 single protons, and a He^{3} nucleus, leaving behind a F^{18} nucleus. It goes without saying that the particles emitted in the first step tend to be more energetic and less isotropic in the laboratory frame than those emitted in the second step. The first step is variously referred to as the "fast" stage, the "knock-on" stage, or more commonly, the "cascade" stage of the spallation reaction, the allusion being to the succession of self-multiplying collisions occurring in the target nucleus, especially if it is quite heavy. The second of the two steps is said to be the "slow" stage, the "de-excitation" stage, or simply the "evaporation" stage. The interested reader will find more detailed discussions of spallation reactions in one of the standard texts on nuclear chemistry, such as that by Friedlander, Kennedy, and Miller.[2]

A problem that has plagued spallation research for years has to do with the emission of "fragments", i.e., clusters of nucleons whose mass number is much lower than that of the target. While most of the fragments appear to be emitted by evaporation, a small fraction of them exhibit energy and angle distributions that are inconsistent with their being emitted by that process. The question then is: at what stage are these rare, anomalous fragments emitted and by what mechanism? Nine years ago, in my introduction to the proceedings of the 1967 conference on spallation[3], this was already noted as an unsolved problem. A great deal of new data on fragments have since become available, but the answer still eludes us. It is possible that the standard two-step model will have to be modified in order to account for these anomalous fragments. Chapter 2 by Hudis includes a discussion of this persisting problem.

So far we have only dealt with spallation reactions in which the incident particle is a single nucleon or pion. The case of two complex nuclei colliding at high energy, i.e. the case of a nucleus-nucleus spallation reaction, can also be described in terms of the Serber model, except that the details are more complex because of the complexity of the incident particle itself. The introduction to the 1967 volume[3] contained a wishful discussion of possible nucleus-nucleus spallation experiments and made an appeal for the acceleration of heavy ions to high energies in existing accelerators. As it turned out, high-energy heavy ions became available sooner than most of us dared expect. In August 1971, both the Berkeley Bevatron and the now-defunct Princeton-Pennsylvania Accelerator succeeded in accelerating usable beams of high-energy nitrogen ions. As a result, interest in nucleus-nucleus spallation multipled, and a new area of research and application, particularly to cancer radiotherapy, was inaugurated. Hudis's Chapter 2 contains a review of recent heavy-ion experiments; Chapter 8 by Chatterjee, Tobias, and Lyman, discusses the depth dosimetry of heavy ions in radiotherapy and diagnosis.

Much of the research on spallation reactions consists in experiments designed to measure cross-sections and calculations designed to predict cross-sections. The early experiments using cosmic rays as incident particles and nuclear emulsion as detectors have now been supplemented by extensive experiments done at high-energy accelerators. In the latter work, measurements range from the counting of radioactive spallation products, to the mass-spectrometric determination of stable product species, to the recent wholesale determination of the mass, charge, energy, and angle distributions of fast emitted particles using scattering chambers and counter arrays. In Chapter 2 Hudis reviews the experimental results on spallation reactions that have become available since the review by Miller in the 1967 volume.[4] In the second half of Chapter 5, Raisbeck and Yiou discuss the type of cross-section data needed in cosmic-ray research and the methods for obtaining them, and, in an Appendix, they give a complete tabulation of the relevant cross-sections measured at the Laboratoire René Bernas.

Because of the complexity of spallation reactions, theoretical research has largely been devoted to Monte-Carlo computer simulations of the physical process. Thus, starting from the early 1950's, a succession of Monte-Carlo programs,

almost all based on some version of the basic Serber model, have
been devised that yielded results of varying degrees of reliability
and illumination. The aim of doing these calculations is either
to test this or that assumption about the reaction mechanism or
to generate reliable values of as yet unmeasured cross-sections
for use in astrophysics, radiation dosimetry, etc. In practice,
these two aims, one physical and the other purely utilitarian,
are not always compatible with one another and are consequently
not always simultaneously attained in a single calculational
model. In Chapter 3, Bertini reviews the various spallation calculations now available or under development.

A special case in the quest for reliable cross-sections is
the so-called "semi-empirical formulas" first devised by Rudstam[5]
in the mid-1950's and recently much extended and improved by
Silberberg and Tsao.[6] These formulas are essentially sophisticated interpolation formulas that interpolate among measured values
of nuclide production cross-sections. Within certain limitations,
the recent Silberberg-Tsao formula has become the most reliable
tool currently available for predicting nuclide production cross-
sections that have not been measured or can not be measured at
accelerators. In the first half of Chapter 4, Silberberg, Tsao,
and Shapiro review this semi-empirical approach and, in the second
half of their chapter, apply the method to cosmic ray problems.
Sections 2 and 5 of Bertini's Chapter 3 and Section 5.7 of
Raisbeck and Yiou's Chapter 5 also discuss this approach.

2.   NUCLEAR CASCADE IN THICK MEDIA

Very often, in situations where spallation reactions are
important, the target in question is not a single nucleus, nor
even a "thin" target that permits no secondary interactions, but
rather a "thick" one in which secondaries from the first interaction can induce further interactions within the target, giving
rise to further secondaries, and so on. When that happens, we
have a "nuclear cascade" in the thick target. Astronomers might
find it suggestive to think of the following atomic analogy: if
we compare individual spallation reactions to the elementary
processes of photon absorption and emission by atoms, then the
nuclear cascade may be compared to radiative transfer.

Thick targets abound in nature: a meteoroid, the Moon's
surface, the Earth's atmosphere, a radiotherapy patient, an
astronaut. The passage, or transport, or cosmic rays through
the Earth's atmosphere offers a good illustration of the nuclear
cascade. When a primary cosmic-ray particle impinges upon the
atmosphere, a chain of self-multiplying spallation reactions is
induced, accompanied by a host of other lower-energy nuclear and
electromagnetic processes. The opulence and the depth of

penetration of this nuclear cascade depends on the primary's energy; eventually the cascade peters out when the energy is shared among a large number of secondaries in the atmosphere. Thus, solar cosmic rays, which have relatively low energy, induce small cascades that take place almost entirely in the top layers of the atmosphere. On the other hand, average galactic cosmic-ray protons several GeV in energy induce nuclear cascades whose remnants are the cosmic radiation we constantly experience at sea-level. Finally, the largest nuclear cascades in the atmosphere are the rare "extensive air showers" or "extensive Auger showers", induced by primaries of energy exceeding 10^{12}eV.

In order to distinguish the nuclear cascade in a thick target from the cascade stage of the two-step Serber spallation model discussed earlier, the latter is sometimes called the "intra-nuclear cascade" stage of the spallation reaction. The entire nuclear cascade in a thick medium — including the intranuclear cascade, the evaporation, other accompanying lower-energy nuclear and electromagnetic processes, and the motion of particles and photons from one interaction to another — can be simulated by the Monte-Carlo method on a computer. In Chapter 7, entitled "Nucleon-Meson Transport Calculations", Alsmiller reviews just this type of nuclear-cascade calculations and compares them with accelerator experiments.

3.   OCCURRENCES OF SPALLATION REACTIONS

Spallation reactions will obviously take place wherever a flux of high-energy particles co-exists with some "stationary" matter with which it can efficiently collide, provided that one or the other or both contain complex nuclei. It is not difficult to think of physical situations where this condition is satisfied, for there are only two main types of high-energy particles in the universe: cosmic rays and particles from man-made accelerators.

Take cosmic rays first. If cosmic rays are accelerated in discrete source objects, then they probably will first of all suffer collisions with the matter in or near the source object. After leaving the source region, cosmic rays will collide with the interstellar (and even intergalactic) matter they encounter while propagating for long periods of time under the influence of magnetic fields. All these are relatively thin targets. Only a minute fraction of cosmic rays will in their lifetime encounter a really thick target such as a star, the Earth, the Moon, a meteoroid, a spacaraft, an astronaut. Those that collide with the Moon or with meteoroids will induce nuclear cascades in the surface layers of these objects. Those encountering the Earth will collide with its atmosphere. Within the atmosphere, the secondary cosmic rays produced in the atmospheric nuclear cascade

might collide with a high-flying aircraft and their occupants.
Fortunately for us, very few strongly-interacting secondaries of
sufficient energy are able to survive to ground level to cause
spallation reactions there. Chapter 9 by Lingenfelter and
Section 3.1 of Alsmiller's Chapter 7 deal with the nuclear cascade in the Earth's atmosphere. Chapter 10 by Davis discusses
the interaction of cosmic rays with the Moon's surface. The
interaction between cosmic rays and meteoroids is not covered
in the present volume, but was treated in the 1967 volume by
Kohman and Bender,[7] who also gave an extensive review of the
nuclear cascade in solid media. The shielding and dosimetry of
cosmic rays in astronautics and aeronautics constituted one of
the earliest practical applications of spallation reactions and
the nuclear cascade; although not covered in the present volume,
the essentials of the problem can still be gleaned from an
earlier review by Shen.[8]

Let us revert for a moment to the collision between cosmic
rays and the interstellar matter. Both contain protons as well as
some complex nuclei. If the cosmic-ray particle is a complex
nucleus, it will undergo spallation and break up. Or, the interstellar gas particle will undergo spallation and break up if it
is a complex nucleus. These two processes, the spallation of
cosmic-ray nuclei and the spallation of interstellar nuclei, are
kinematically equivalent and occur with comparable frequency.
Yet, their consequences in the laboratory frame are quite different.
In the first process, the breakup fragments of the cosmic-ray
nucleus will inherit the cosmic-ray velocity and thus have a
chance to be observed at the Earth as cosmic rays. In the second
process, the breakup fragments of the interstellar nucleus will
in general move off slowly in the laboratory frame and will be
thermalized by electromagnetic processes to become a part of the
interstellar gas. The first process has long been the subject of
a great deal of cosmic-ray research; it is this process that has
traditionally been called "fragmentation" by cosmic-ray physicists
and that has successfully explained the overabundance of Li, Be,
and B observed in the cosmic radiation.

The second process, the spallation of interstellar matter,
was practically ignored until 1961 when Milford and Shen[9,10]
first called attention to its astrophysical importance in the
production of Li, Be, and B in interstellar matter; they suggested,
on the basis of production rates they gave, that most interstellar
Li, Be, and B, and perhaps also H^2, were produced by this mechanism
rather than by stellar-surface spallation as then often believed.[10]
They also pointed out that the observed abundances of Li, Be, and
B could serve as an "integrating meter" for interstellar cosmic
rays of energy as low as ~50 MeV, and they showed that the then
observed upper limits on the interstellar Be abundance and on the
Li abundance in the nebula of T-Tauri placed an upper limit of

~1 to ~10 $cm^{-2}sec^{-1}$ on the average intensity of such cosmic rays over the past few billion years.[9] Almost a decade later, this method of manufacturing Li, Be, and B and of deriving the past cosmic-ray intensity was independently rediscovered by Reeves, Fowler, and Hoyle[11] and developed by them and by others into a new coherent theory of the origin of the light elements. Chapter 4 (second half) by Silberberg, Tsao, and Shapiro and Chapter 5 by Raisbeck and Yiou deal primarily with the first process, while Chapter 6 by Audouze, Meneguzzi, and Reeves deals with both the first and second processes.

We now turn to the other type of high-energy particles, those from accelerators. Accelerated particles will collide with experimental targets, absorbers, measuring instruments, shielding walls, radiobiological specimens, radiotherapy patients. Some of these targets are "thin", but most are "thick". The shielding wall and the ionization spectrometer are considered in Sections 3.2 and 3.3, respectively, of Alsmiller's Chapter 7. The case of the radiotherapy patient irradiated by heavy ions is taken up in Chapter 8 by Chatterjee, Tobias, and Lyman. The main advantage of heavy ions in radiotherapy is the high spatial rate of energy deposition that occurs at the end of their ionization tracks. An important item of information needed in the actual use and evaluation of heavy-ion radiotherapy is the distribution of dose within the irradiated tissue, the so-called depth-dose distribution. The human body is thick enough compared with the nuclear interaction mean-free-path for some of the incident ions to induce spallation reactions, thereby spoiling the simple depth-dose distribution that would otherwise obtain if atomic processes alone were operative. The detailed calculation of the depth-dose distribution can of course be handled just like that of any nuclear cascade, once a satisfactory method for calculating nucleus-nucleus spallations is available. The RENO spallation model of the University of Pennsylvania,[12] referred to in Section 4 of Bertini's Chapter 3, can now handle nucleus-nucleus spallations in which the incident energy is as high as a few GeV per nucleon and in which the mass numbers of the interacting nuclei add up to as high as 33, sufficient for the case of nitrogen ions on tissue. It is hoped that the next few years will see the application of heavy-ion spallation models such as this to the detailed calculation of the depth-dose distribution of incident heavy ions so that the full potential of heavy-ion radiotherapy can be exploited. Few other applications of spallation nuclear reactions bear so directly on man.

REFERENCES

1.   R. Serber, Phys. Rev. 72, 1114 (1947)

2.   G. Friedlander, J. W. Kennedy, and J. M. Miller, Nuclear and Radiochemistry, 2nd ed. (New York, John Wiley & Sons, Inc., 1964), esp. Chapter 10.

3.   B. S. P. Shen in High-Energy Nuclear Reactions in Astrophysics, edited by B. S. P. Shen (New York: W. A. Benjamin, Inc., 1967), p. 1.

4.   J. M. Miller in High-Energy Nuclear Reactions in Astrophysics, edited by B. S. P. Shen (New York: W. A. Benjamin, Inc., 1967), p. 19.

5.   G. Rudstam, Phil. Mag. 46, 344 (1955).

6.   R. Silberberg and C. H. Tsao, Astrophys. J. Suppl. 25, 315 and 335 (1973).

7.   T. P. Kohman and M. L. Bender, in High Energy Nuclear Reactions in Astrophysics, ed. by B. S. P. Shen (New York: W. A. Benjamin, Inc., 1967), p. 169.

8.   B. S. P. Shen, Astronautica Acta 9, 211 (1963).

9.   S. N. Milford and B. S. P. Shen (S. P. Shen), Phys. Rev. 122, 1921 (1961).

10.   S. N. Milford and B. S. P. Shen (S. P. Shen), Bull. Am. Phys. Soc. 7, 62 (1962).

11.   H. Reeves, W. A. Fowler, and F. Hoyle, Nature 226, 727 (1970).

12.   W. F. Schmitt, C. L. Ayres, M. Merker, and B. S. P. Shen, Proc. XIII Int'l Cosmic-Ray Conf. (University of Denver, 1973) p. 517; C. L. Ayres, W. F. Schmitt, M. Merker, and B. S. P. Shen, Ibid., p. 522.

J. Hudis

Chemistry Department, Brookhaven National Laboratory,
Upton, New York 11973

There has been a long if sometimes distant relationship
between chemists who study high energy nuclear reactions and
astrophysicists. Some of the results of these nuclear reaction
studies, usually product yield cross sections for a variety of
target nuclei, are required by the astrophysics community to
determine the composition, energy distribution and intensity of
both cosmic and galactic radiation. However results of experiments
designed to elucidate high energy reaction mechanisms are
frequently not directly useful for these ends. The paper by
J. M. Miller (1967) in the previous volume of this series reviewed
the models and calculations which have proven useful in understanding high energy reaction data and compared the results of those
calculations with experiment. The following chapter in this
volume by H. Bertini (1976) continues that effort; here we concentrate on describing experimental results obtained subsequent to
1967 and discuss their significance.

Nuclear chemists like to work with target nuclei seldom
lighter than vanadium, whereas the interesting and relatively
abundant target materials in space are chiefly carbon to silicon.
In part this preference for heavy targets comes about because up
until the past few years chemists usually limited themselves to
the measurement of radioactive products and one quickly runs out
of candidates with reasonable half lives when targets lighter than
iron or copper are irradiated. However, this picture has changed
considerably in the last five years and this limitation no longer
holds for many investigations at a number of laboratories. A
second difficulty is that nuclear chemists have favorite elements
which they always investigate and these usually do not correspond
to the elements of greatest interest to astrophysicists. For

many reasons, one of which is historical, copper is a favorite
element but iron and nickel are not. Finally most nuclear
reaction studies are carried out with thin metallic foils to
ensure that the observed products result from reactions between
nuclei of a single element and incident particles of well defined
energy. Many samples of astrophysical interest on the other hand
are thick and are composed of a number of different elements.
Thus the measured radioactive species are products of nuclear
reactions in a sample in which a number of elements have been
irradiated both by a primary particle and a spectrum of secondaries.

In spite of these difficulties the astrophysics community has
been able to obtain most of its needed reaction data. Probably
the most useful tool for this purpose has been the semi-empirical
formulae of Rudstam (1966), Silberberg and Tsao (1973), as well
as others, which do a good job predicting yields of products from
different target nuclei irradiated by energetic particles. This
is especially true for targets not much lighter than vanadium,
for products not too far removed from beta stability, and for
protons as incident particles.

In this chapter the problems, questions, and available
facilities which provide the impetus for most of the recent
efforts in high energy spallation reaction research will be
reviewed. The author's personal prejudices, national chauvenism,
and ignorance of some on-going relevant programs will undoubtedly
be revealed and for this I apologize in advance. No attempt at a
comprehensive literature survey was made and only one or two
recent references to various on-going projects are cited.

One major difficulty in discussing high energy nuclear
reactions is terminology. For the basis of this review spallation
is defined as everything but fission. Fission in turn is defined
as the disintegration of a nucleus into two or more relatively
equal size fragments. For organizational reasons this survey
groups experiments according to projectile, and results which may
be of special interest to members of the astrophysical community
will be noted.

1. PROTONS

1.1 ≤500 MeV

Most of the spallation studies in this energy range, especially
those using light targets such as carbon, nitrogen and oxygen,
are specifically aimed at understanding astrophysical problems.
One should mention at least the emulsion work at Strasbourg
(Jung et al, 1970), and the large amount of information obtained
by mass spectrometric and radiochemical techniques at the Orsay
laboratory (Epherre and Iseide, 1971). However, mention should

also be made of more recent work which is being done by nuclear
physicists and chemists using scattering chamber and counter
techniques. In at least three laboratories in the United States,
Michigan State University (Panggabean et al, 1974) and the Uni-
versities of Maryland (Roche et al, 1975) and Washington (Oberg
et al, 1975), detailed measurements of excitation functions, and
energy and angular distributions of light fragments such as
lithium, beryllium and boron produced from light targets such as
carbon, nitrogen, oxygen and neon are being made with this
technique.

1.2   0.5 to 30 GeV

There are relatively few spallation cross section studies taking
place in this energy region. This is true even though there are
still large gaps in our knowledge about spallation yields 20-30
mass numbers away from the target from heavy target nuclides.
Most of the cross section work in this field now, at least below
30 GeV, is carried out mainly to compare with pion, heavy ion and
300-GeV proton results. Included in such work are the 350-MeV
(Hudis et al, 1975) and 3.9-GeV (Cumming et al, 1974) proton
spallation of copper and the 11.5-GeV proton spallation of silver
(English et al, 1974), gold (Weisfield et al, 1975) and uranium
(Yu et al, 1975). Most of the present effort is aimed at answer-
ing two major questions:

(1)   What process or combination of processes is responsible
for the complex distribution of nuclides observed in the middle
mass region when heavy target nuclides are irradiated by GeV
projectiles?

(2)   What is the mechanism by which light fragments, whose
mass may reach as high as 30 or 40, are produced from heavy
elements irradiated with GeV particles?

Cross section measurements even at 1 and 2 GeV showed that
in addition to a so-called normal fission product yield consisting
mainly of neutron-excess species, there were appreciable amounts
of neutron-deficient species and the relative amounts of the
neutron-deficient species increased with increasing bombarding
energy.

Figure 1 from the work of Chu et al (1971) indicates the
very complex distribution of product yields obtained when uranium
is irradiated by GeV protons. It is obvious, especially from the
mass 131 distribution in Fig. 1, that cross section measurements
alone cannot disentangle and elucidate the mechanism or combination
of mechanisms responsible for these yields. Thus recoil experiments
in which the kinetic energy of specific product nuclides are
measured, sometimes as a function of angle, have played an

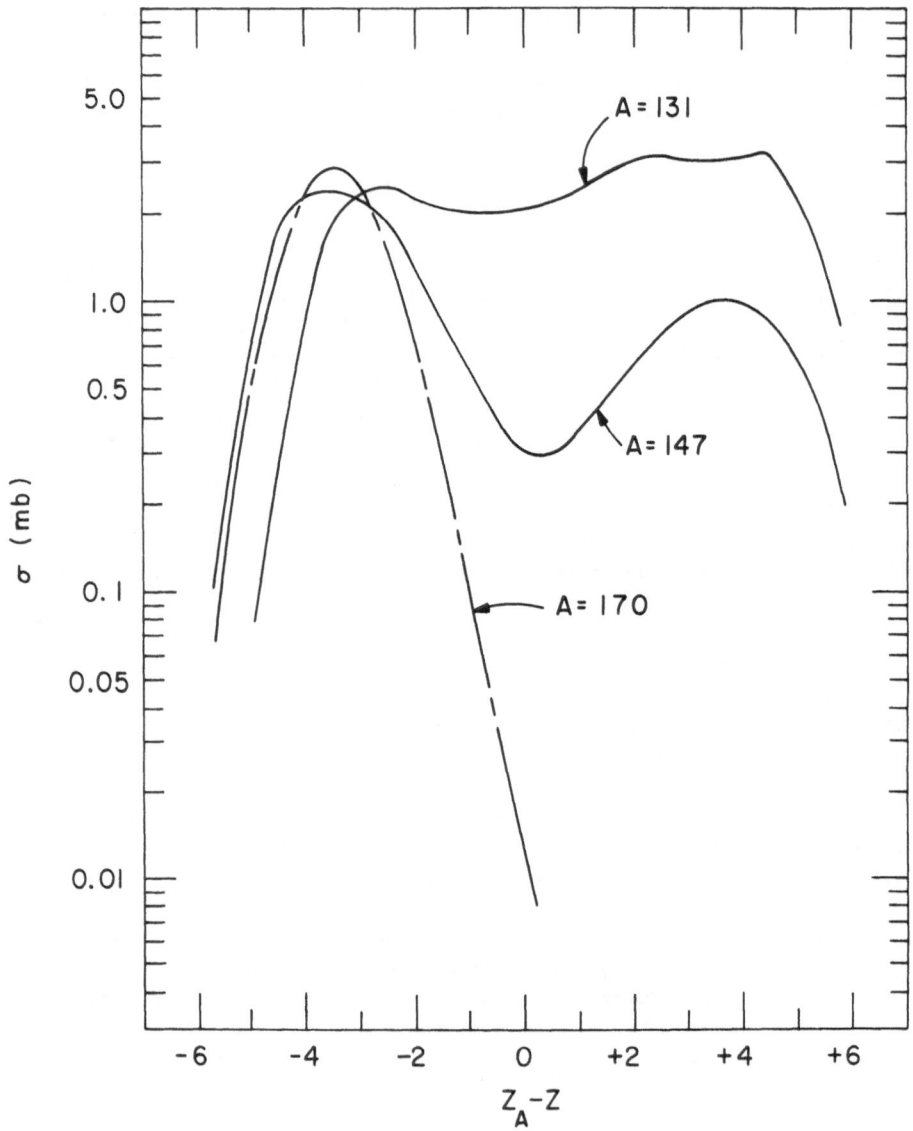

Fig. 1. Distribution of cross section with the products' nuclear charge for the mass number regions at about A = 131, 147, and 170 observed in the 28–GeV proton irradiation of U (Chu et al, 1971). Z_A – Z = 0 defines the most stable charge for each mass number.

increasing role in the study of these processes. Analysis of
both cross section and recoil data (Cumming and Bächmann, 1972;
Beg and Porile, 1971) show that the neutron-excess side of the
distribution is due to a fission process similar to that observed
at much lower energies. The neutron-deficient products have very
low momenta and thus resemble spallation products. However the
mean forward momentum transfer associated with their production
is much lower than that estimated for a "classical" spallation
process and the ejection of light fragments in the early stages
of these high energy interactions has been suggested as a possible
solution to this problem.

The study of these light fragments has been of interest since
the very earliest experiments in the GeV region. It was obvious
then and remains obvious now that the production of such fragments
which have energy thresholds of 100's of MeV is definitely a high
energy process and in addition it is one which cannot be explained
easily by the theories and models which go so far in correlating
almost all of the other high energy data. For this reason, of
course, the study of these fragmentation products is especially
interesting because hopefully one may be able to learn new facts
about nuclear reactions or about the properties of nuclei at high
temperatures. The scattering chamber experiments at Argonne
National Laboratory, Brookhaven National Laboratory, Los Alamos
Scientific Laboratory and Lawrence Berkeley Laboratory constitute
the major effort by U.S. nuclear chemists towards the study of
these fragments. Angular and energy distributions of emitted
fragments are measured in all of these experiments but the thrust
of each program is different.

In the LBL experiments, the relative yields of single fragments with $Z \leq 20$ from uranium (Poskanzer et al, 1971) and $Z \leq 14$
from silver (Hyde et al, 1971) produced in 5.5-GeV proton irradiations were measured. Figure 2 shows the clean separation of
adjacent elements achieved up to $Z \sim 18$ from U and the decrease
of yield with increasing charge. The energy distribution of each
element was determined at a number of angles and the spectra of
^{4}He, ^{7}Li, and ^{10}Be from U targets are shown in Fig. 3. The energy
distributions look similar to what one would expect for evaporation
processes, that is a Maxwellian distribution, but detailed analysis
showed many points where this simple model fails. Thus the energy
spectrum of fragments extends to low enough energies that Coulomb
barriers equal to only one-half the classical value are required.
In addition, no single nuclear temperature can fit the whole
spectrum, but one finds temperatures of about 10-13 MeV near the
peak of the distribution and values as high as 20 MeV for the
high energy tail. Even temperatures of 10 MeV indicate that if
equilibrated there is sufficient excitation energy to dissociate
the nucleus completely into individual nucleons. Finally, the

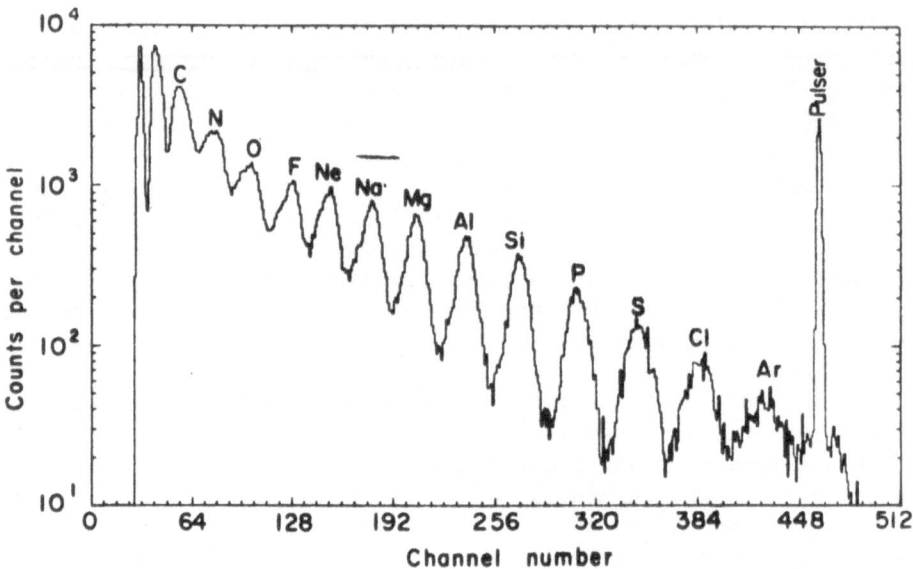

Fig. 2.   Particle spectrum obtained by Poskanzer et al (1971) in
the 5.5-GeV proton irradiation of U.

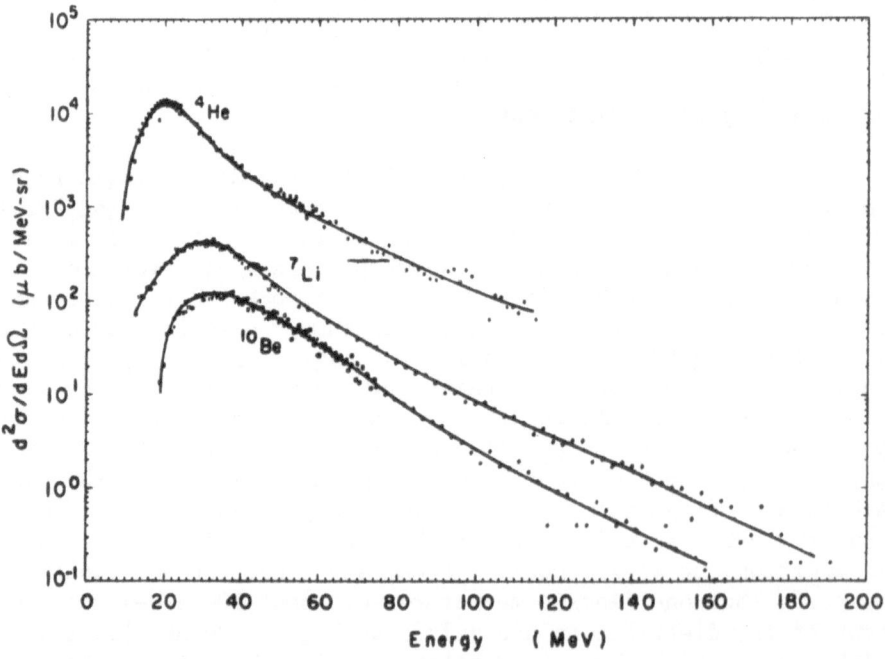

Fig. 3.   Kinetic energy spectra of three fragments emitted in the
5.5-GeV proton irradiation of U (Poskanzer et al, 1971).

observed forward peaking high energy fragments cannot be explained on the basis of the two-step model so successful for high energy nuclear reactions in general. It should be pointed out that only about 10% of the total fragment yield lies in this unexplained region. Whether minor but still unknown modifications to the cascade-evaporation model can bring about better agreement between calculation and experiment remains to be seen. New and uniquely high energy mechanisms may still be required. Similar experiments are now in progress at LAMPF with 800-MeV protons as the incident projectiles (Butler and Perry, 1975).

At BNL similar experiments with uranium and gold plus 28-GeV protons were carried out. Probably the most interesting result to date is shown in Fig. 4 : peak yields of fragments with Z between 6 and 20 come not at 0° to the beam, as at 5.5 GeV, but at about 70° (Remsberg and Perry, 1975). The question is, does the mechanism for the production of these fragments change between 5.5 and 28 GeV, or will the correct mechanism describing the lower energy data incorporate this shift in preferred emission angle at the higher energy? These results as well as recent findings in heavy-ion induced reactions have revived interest in some old ideas using a hydrodynamic model of nuclear reactions (Glassgold et al, 1959). Shock waves traveling through nuclear matter would lead to fragment emission at angles predicted to be very close to those observed by Remsberg and Perry (1975). Another explanation comes from some recent theoretical results which at even higher energies predicts copious production of pions along the beam trajectory and perhaps such processes will simply not allow zero degree production of fragments. A second BNL experiment in which light fragments have been measured in coincidence with another fragment has recently been completed but the data are just now being analyzed. Anything we can learn about the nature of the partners of these light fragments will certainly give us a better picture of their production mechanism. The general belief is that these light fragments are the partners of the very neutron-deficient spallation products discussed above. This belief may arise, in part, from the desire to solve two long-standing problems at the same time.

The third such experiment, at ANL, studies the reaction between 11.5-GeV protons and gold (Wilkins et al, 1975). The energy spectra of helium, lithium, and beryllium nuclides were measured in coincidence with fission fragments at a number of angles. This study is specifically aimed at determining if the light fragments are emitted from the struck target nucleus or from a highly excited fission fragment.

A recent experiment at LBL is perhaps of greater direct interest to astrophysicists. Here, C and Al targets were irradiated with 2.1- and 8.4-GeV protons and the energy spectrum

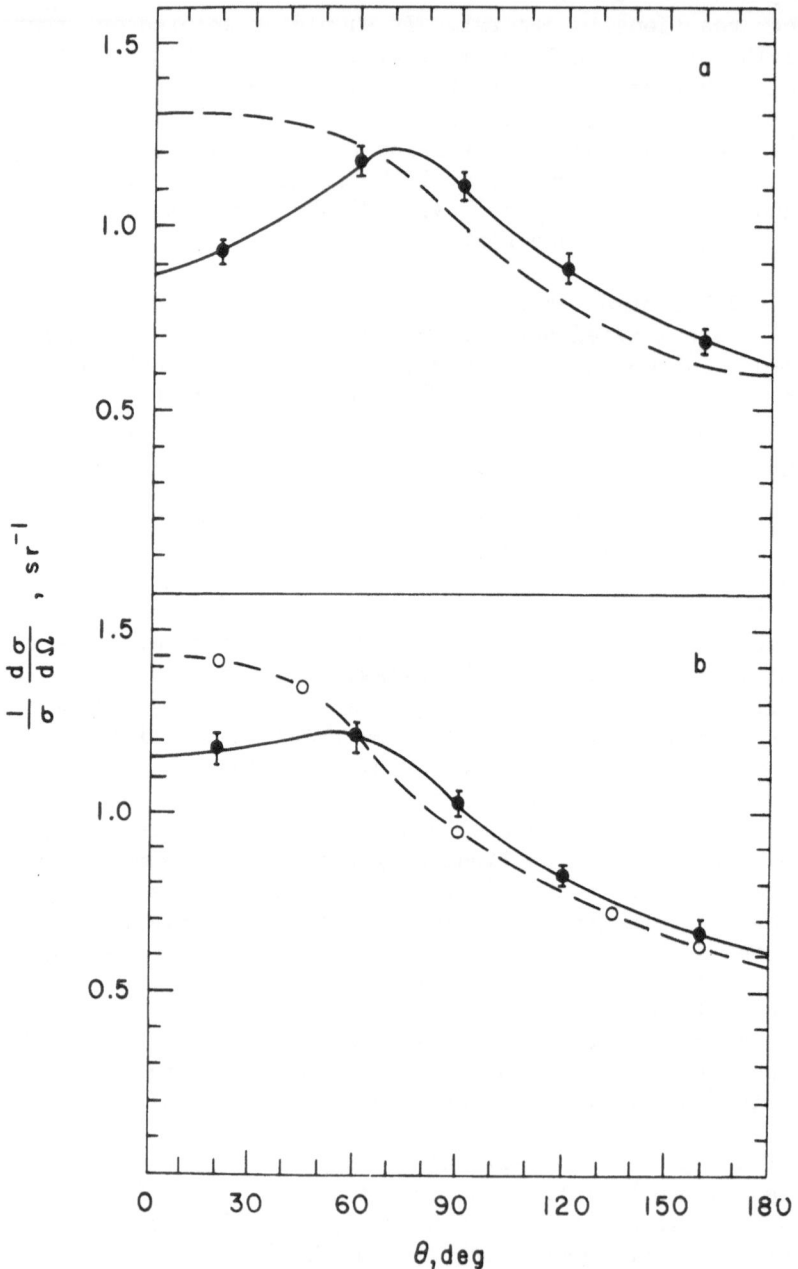

Fig. 4.   (a) Angular distribution of Na fragments from U + 28-GeV protons, solid curve, compared to the distribution of ^{24}Na fragments from Bi + 2.9-GeV protons, dashed curve.
(b) Angular distribution of C fragments from U + 28-GeV protons, solid curve, and from U + 5.5-GeV protons, dashed curve (from Remsberg and Perry, 1975).

of each product element, up to C for C targets, and up to Mg for
Al targets, was measured at 3-5 angles (Sextro et al, 1975). We
shall refer to this work again.

1.3　300–400 GeV

A large number of fission and spallation studies have been
completed or are in progress at the Fermi National Accelerator
Laboratory. Spallation yield distributions from vanadium (Katcoff
et al, 1973), cobalt (Katcoff et al, 1973), and silver (English
et al, 1974) and the yields of some spallation products from gold
(Weisfield et al, 1975) and uranium (Chang and Sugarman, 1974;
Yu and Porile, 1975) have been measured. The zeroeth order finding
in all of these experiments is that spallation cross sections
change very little between 11.5 and 300 GeV and possibly not at
all between 30 and 300. There are, however, real differences in
the recoil properties of the spallation products between 11.5
and 300 GeV and work continues to understand these differences in
terms of what is known about elementary particle interactions at
these energies. Thus, as has been assumed all along, the use of
spallation cross sections obtained at 10-30 GeV is probably a
very good approximation for all higher proton energies.

2.　PIONS

Pion beams of high intensity are available at the Los Alamos Meson
Physics Facility (LAMPF), and the SIN facility in Switzerland is
just coming into operation. The incident pion energies available
range up to 400 to 500 MeV. Since the production and reabsorption
of pions in proton-induced reactions has been postulated as a most
important mechanism for the conversion of projectile kinetic
energy to nuclear excitation energy many nuclear chemistry studies
are under way or planned on those accelerators.

Many of these experiments are designed to test in a critical
way many of the assumptions and input parameters of the nucleonic
cascade calculations which are discussed in Dr. Bertini's chapter
in this volume. Thus experiments are under way to determine:

(1)　just how pions interact with complex nuclei;

(2)　if the models and the assumptions used to explain high
energy proton spallation, which is strongly dependent on pion
production and reabsorption processes in the struck nucleus, are
good ones.

One recently completed study was that of the $^{12}C(\pi^{\pm},\pi^{\pm},n)^{11}C$
reaction from 50 MeV to 550 MeV (Dropesky et al, 1975). Since

this is such a simple nuclear reaction on a light nucleus one
would expect that the behavior of the simple elementary particle
interactions responsible for the observed product would be
mirrored in this reaction. Indeed such effects have been
observed, the peaks in the cross section follow the peaks in the
elementary particle cross sections but interestingly not exactly.
The (π^+,π^+n) resonance peak for the reaction occurs about 20 MeV
lower and that for the (π^-,π^-n) reaction about 20 MeV higher than
the 185-MeV value observed in the elementary particle interactions.
In addition, calculations of Sternheim and Silbar (1975) based
simply on the relative probabilities of all the possible elementary
particle reactions which could lead to the product are in good
agreement with the experimental data. This itself is no small
feat since previous attempts to explain similar data obtained
with high energy protons have been unsuccessful.

Spallation studies on aluminum and copper are in progress.
Early results, comparing 195-MeV π^+,π^- data with 350-MeV proton
data show no great surprises (Hudis et al, 1975). The effects of
the charge difference between π^+ and π^- are seen in spallation
products quite far from the target nucleus. The proton energy
of 350 MeV was chosen to compare results at the same total energy.
Preliminary results indicate that higher cross sections are
obtained with pions for spallation products far from the nucleus
than with protons. This indicates that more complex cascades are
initiated by direct pion absorption than by the higher energy
proton. These and similar data will provide good tests for the
Monte Carlo calculations and hopefully allow us to refine and
improve details of the models of the nuclear interactions and
nuclear structure used as starting points.

3. HIGH ENERGY HEAVY IONS

With heavy ion beams of 2-, 4-, and 28-GeV ^{14}N ions at the Princeton Particle Accelerator and LBL Bevatron a new field of nuclear
reaction studies was initiated. More recently the use of the LBL
Super Hilac as the injector for the Bevatron—the Bevalac—has
permitted acceleration of projectiles ranging from ^{2}H to ^{40}Ar at
2 GeV/nucleon and beams of Kr ions are to be tested this year.
Study of heavy ion induced spallation and fission reactions along
with pion induced reactions are now a major interest of nuclear
chemists in this field.

The initial experiments with heavy ions are designed to
answer the following questions:

(1) Are there new phenomena appearing with very energetic
heavy ions?

(2) Are heavy ions more efficient than protons in transfer-
ring energy to target nuclei, and if so, why?

(3) Up to what incident energies can heavy ion results be
explained adequately by the theories which work well for ions at
≤10 MeV/nucleon?

(4) Can the theories developed for high energy nucleon
induced reactions successfully take over this task when the low
energy models fail?

Zebelman et al (1974) at LBL have used the counter technique
described previously to measure single fragment production from U
irradiated with 2.1 GeV/nucleon ^{2}H and ^{4}He ions. Only He, Li, Be,
and B fragments were measured but the results are quite clear —
deuterons are not very different from protons. Although their
cross sections are ∿50% higher, the fragments' energy spectra are
very similar to those obtained with incident protons. However
appreciable differences do show up with alpha particles; cross
sections are 4-5 times larger and the energy spectra are different.
Recall that with 5.5-GeV protons, Coulomb barriers about 50% of
the classical values and nuclear temperatures varying from 10-20
MeV were required to fit the low and high energy portions of the
spectra, respectively. These parameters become even more
"unphysical" in the alpha particle work; the barrier must be
lowered another 15% and the apparent nuclear temperature raised
another 1.5 MeV. Although the usefulness of these parameters is
not at all clear, it is obvious that deposition energies are larger
with alpha particles than with deuterons or protons.

Another fragment study, this time with the Lexan track
detector technique, was used by Sullivan et al (1973) to study
the interaction of 34-GeV ^{16}O ions with Au. Similar conclusions
were reached: compared to 2-GeV protons, the ^{16}O data show a
much higher frequency of fragment ejection from the target nucleus.
The cross section for fragment emission in the ^{16}O irradiations
was ∿12 times that measured in the proton irradiation, but the
figure varies with the mass of the fragment, being ≈25 for Z = 8
and ≈3 for Z ≳ 12. The integrated cross section for fragment
production is higher than geometric indicating multiple emission.
The fragments have low kinetic energies, requiring Coulomb
barriers ∿1/3 that of the classical value much like the result
obtained with 8-GeV alpha particles on U. No evidence was seen
of new reaction phenomena; in particular there was no indication
of fragments produced by shock waves as predicted by Glassgold,
Heckrotte, and Watson (1959).

Positive evidence for shock waves has been claimed by
Schopper (1975) who used AgCl plates, not emulsions, to look at
stars and fragments produced by GeV/nucleon C and O ions. He

found that the angular distributions of emitted high energy
protons and alpha particles peak at forward angles and that the
peak position varied with incident energy in a way that rules out
nucleonic cascade and evaporation mechanisms as possible sources
of these particles. These heavy ion results along with the high
energy proton results of Remsberg and Perry (1975) mentioned
previously are the only ones known to this reviewer which at
least tentatively support shock waves as a possible nuclear
reaction mechanism. A good deal of work, both experimental
and theoretical, is now going on in this field.

A more typical radiochemical type of experiment has been
reported by Cumming et al (1974) in which the spallation yields
from Cu + 3.9-GeV ^{14}N ions and 3.9-GeV protons were measured.
Absolute cross sections for the heavy ion experiment were not
obtained but it was shown that both the shapes of the mass yield
curve and of the charge distribution curves were very similar with
both projectiles. As is shown in Fig. 5, except for products of
secondary reactions (^{62}Zn, ^{66}Ga and ^{69}Ge) and the two lightest
nuclides measured (^{3}H and ^{7}Be), the cross section ratios are very
close to 1.0 with no obvious difference between neutron-excess
and neutron-deficient nuclides. This finding is somewhat un-
expected since the nitrogen energy is only ∿280 MeV/nucleon, well
below the energy where pion processes become important. A new
experiment with 2-GeV/nucleon ^{12}C (Cumming and Haustein, 1975)
is in progress to see if drastic changes in the yield patterns
appear, similar to those observed for protons of between 280 MeV
and 2 GeV. In addition similar studies with 25-GeV ^{12}C ions + Ag
(Rudy and Porile, 1975) and U (Otto et al, 1975) have been
reported.

The last heavy ion research to be discussed are the experiments performed by the Heckman group at the Bevalac (Heckman et al,
1975; Lindstrom et al, 1975). A series of targets ranging from
protons to lead were irradiated with GeV/nucleon beams of ^{12}C,
^{14}N, and ^{16}O and the spectra of fragments emitted at 0° to the
beams measured. They have made the following interesting observations:

(1) A very large fraction of the emitted fragments with
Z ≥ 2 were emitted at 0° and their spectra showed that minimal
amounts of momenta, both in and perpendicular to the beam
direction, had been transferred during the interaction.

(2) The distribution of fragments was independent of the
target nucleus.

(3) Their measured fragment cross sections from ^{12}C and ^{16}O
incident on protons were about equal to those in the literature
for the reverse reaction, protons incident on ^{12}C and ^{16}O. The

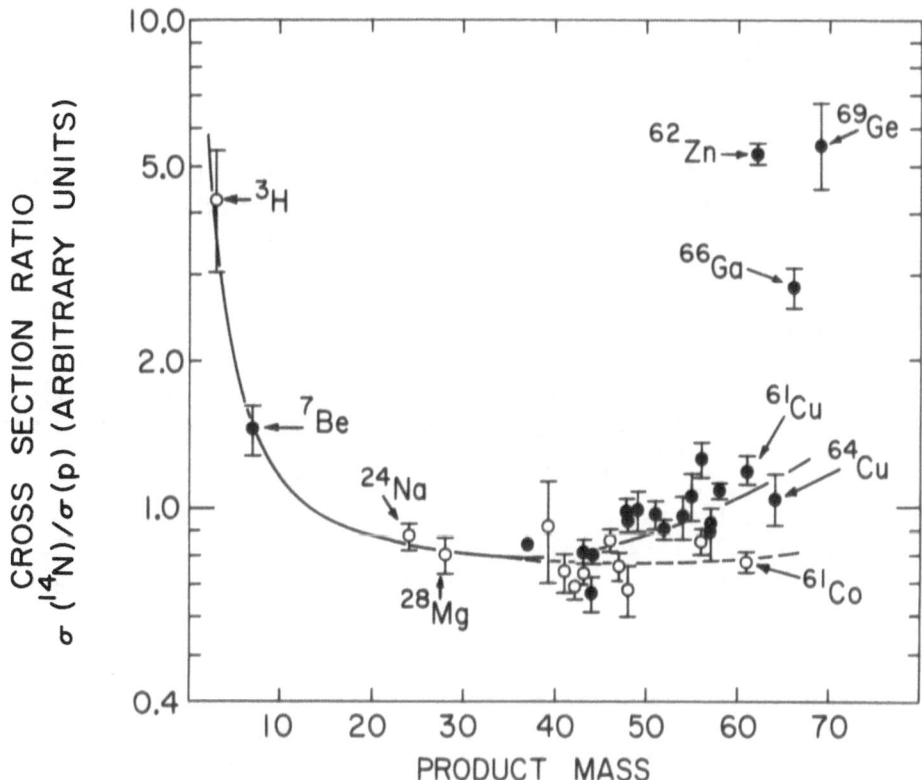

Fig. 5. Ratio of yields of spallation products of Cu irradiated with 3.9-GeV protons and ^{14}N ions. The open circles are neutron-excess products, the closed circles neutron-deficient ones (from Cumming et al, 1974).

recent work by Sextro et al (1975) where fragment yields from C
and Al targets irradiated by 2.1-GeV protons were measured was
carried out to check more directly this interesting result.

(4)   The same fragment cross sections were obtained with
2-GeV/nucleon and 1-GeV/nucleon ^{12}C ions.

(5)   The yield of a particular fragment from a given target
nucleus was relatively independent of projectile type. The fact
that these ratios are not very far from 1, actually 40% were
1.0 ± 15%, is surprising since the actual cross sections varied
by a factor of 1000.

To these observations we add two more:

(6) $\sigma(^{14}N + Cu \rightarrow$ spal. prod.$)/\sigma(^{1}H + Cu \rightarrow$ spal. prod.$) \cong$
constant, from the Cumming et al (1974) experiment performed with
4-GeV ^{14}N ions and 4-GeV protons.

(7) $\sigma(^{1}H +$ target $\rightarrow$ spal. prod.$)$ is energy independent above
10 GeV for all targets (Weisfield et al, 1975; Katcoff et al,
1973; English et al, 1974; Chang and Sugarman, 1974; Yu et al,
1975).

Heckman et al (1972) have applied to their data some approaches
introduced by elementary particle theorists (Frazer et al, 1972)
to explain multiparticle reactions at very high energies. These
theories allow only minimal momentum transfer between beam and
target nuclei. Since minimal momentum transfer was observed in
the beam fragments, it is possible that these ideas may be
applicable in relativistic heavy ion reactions. They are:

Limiting fragmentation (scaling). In a reaction $B + T \rightarrow F + X$,
where B, T, and F are the beam, target, and a product nucleus
and X is everything else, the hypothesis of limiting fragmentation
predicts that at sufficiently high energies the reaction is energy
independent. The same argument holds for reaction products origi-
nating from either the beam or target nucleus. Items 4 and 7,
above, are empirical findings that seem to agree with this
hypothesis. It is interesting that reactions in which either the
target or projectile or both are complex nuclei lends support to
this theory at much lower energies than has been observed with
elementary particle interactions.

Factorization. This hypothesis asserts that the cross section
σ_{BT}^{F} can be factored into separate and independent terms, one of
which depends only on the parent-fragment pair and the other on
the "spectator" nucleus. Thus

$$\sigma^F_{BT} = \gamma^F_B \gamma_T \text{ when F is a fragment from the beam projectile}$$

and

$$\sigma^F_{BT} = \gamma^F_T \gamma_B \text{ when F is a fragment from the target nucleus.}$$

Factorization predicts that at sufficiently high energies, modes of fragmentation of the beam particles will become independent of target (item 2) and conversely that the mode of disintegration of the target nucleus will become independent of the beam projectile (item 6). Thus a number of experimental results seem to support, at least tentatively, these two hypotheses. However some recent work indicates that this model does not hold for all targets, and the factorization relationship may break down for very light targets (Raisbeck and Yiou, 1975).

The ideas discussed above are macroscopic models, reminiscent of phase space arguments, and as such tell us little about the detailed microscopic interactions taking place as is attempted by the nucleonic cascade calculations. It is obvious, however, that if these hypotheses continue to correlate relativistic heavy ion reaction data successfully, their predictive powers will be of great value to the astrophysics community in estimating yields for a large class of nuclear reactions.

ACKNOWLEDGMENT

Work at Brookhaven National Laboratory was performed under the auspices of the U.S. Energy Research and Development Administration.

REFERENCES

Beg, K. and Porile, N. T. 1971, Phys. Rev. C, 3, 1631.

Bertini, H. 1976, this volume.

Butler, G. W. and Perry, D. G. 1975, presented at the 170th ACS National Meeting, Chicago, August 25-28.

Chang, S. K. and Sugarman, N. 1974, Phys. Rev. C, 9, 1138.

Cumming, J. B. and Bächmann, K. 1972, Phys. Rev. C, 6, 1362.

Cumming, J. B. and Haustein, P. E. 1975, private communication.

Cumming, J. B., Haustein, P. E., and Stoenner, R. W. 1974, Phys. Rev. C, 10, 739.

Chu, Y.-Y, Franz, E.-M., Friedlander, G., and Karol, P. J. 1971, Phys. Rev. C, 4, 2209.

Dropesky, B. J., Butler, G. W., Orth, C. J., Williams, R. A., Friedlander, G., Yates, M. A., and Kaufman, S. B. 1975, Phys. Rev. Lett., 34, 821.

English, G., Porile, N., and Steinberg, E. P. 1974, Phys. Rev. C, 10, 2268.

English, G., Yu, Y. W., and Porile, N. T. 1974, Phys. Rev. C, 10, 2281.

Epherre, M. and Iseide, C. 1971, Phys. Rev. C, 3, 2167.

Frazer, W. R., Ingber, L., Mehta, C. H., Poon, C. H., Silverman, D., Stowe, K., Ting, P. D., and Yesian, H. J. 1972, Revs. Modern Phys., 44, 284.

Glassgold, A. E., Heckrotte, W., and Watson, K. M. 1959, Ann. Phys. (New York), 6, 1.

Heckman, H. H., Greiner, D. E., Lindstrom, P. J., and Bieser, F. S. 1972, Phys. Rev. Lett., 28, 926.

Hudis, J., Harp, G. D., Dropesky, B. J., Norris, A. E., Orth, C. J., and Williams, R. A. 1975, presented at the 170th ACS National Meeting, Chicago, August 25-28.

Hyde, E. K., Butler, G. W., and Poskanzer, A. M. 1971, Phys. Rev. C, 4, 1759.

Jung, M., Jacquot, C., Baixeras-Aiguabella, C., Ischmitt, R., and Braun, H. 1970, Phys. Rev. C, 1, 435.

Katcoff, S., Kaufman, S., Steinberg, E. P., Weisfield, M. W., and Wilkins, B. D. 1973, Phys. Rev. C, 30, 1221.

Lindstrom, P. J., Greiner, D. E., Heckman, H. H., Cork, B., and Bieser, F. S. 1975, LBL-3650.

Miller, J. M. 1967, in High-Energy Nuclear Reactions in Astrophysics, ed., B.S.P. Shen, Ch. 1, pp. 19-36 (New York: W. A. Benjamin, Inc.).

Oberg, D. L., Bodansky, D., Chamberlin, D., and Jacogs, W. W. 1975, Phys. Rev. C, 11, 410.

Otto, R. J., Fowler, M. M., Binder, I., Lee, D., and Seaborg, G. T. 1975, presented at the 170th ACS National Meeting, Chicago, August 25-28.

Panggabean, L. M., Austin, S. M., and Laumer, H. 1974, Phys. Rev. C, 10, 4.

Poskanzer, A. M., Butler, G. W., and Hyde, E. K. 1971, Phys. Rev. C, 3, 882.

Raisbeck, G. M. and Yiou, F.   1975, Phys. Rev. Lett., $\underline{35}$, 155.

Remsberg, L. P. and Perry, D.   1975, Phys. Rev. Lett., $\underline{35}$, 361.

Roche, C. T., Clark, R. G., Mathews, C. J., and Viola, V. E. 1975, private communication, V. Viola.

Rudstam, G.   1966, Z. Naturforsch., $\underline{21a}$, 1027.

Rudy, C. R. and Porile, N. T.   1975, presented at the 170th ACS National Meeting, Chicago, August 25–28.

Schopper, E.   1975, private communication.

Sextro, R. G., Zebelman, A. M., and Poskanzer, A. M.   1975, presented at the 170th ACS National Meeting, Chicago, August 25–28.

Silberberg, R. and Tsao, C. H. 1973, Astrophys. J. Suppl. $\underline{25}$, 315 and 335 (1973).

Sternheim, M. M. and Silbar, R. R.   1975, Phys. Rev. Lett., $\underline{34}$, 824.

Sullivan, J. D., Price, P. B., Crawford, H. J., and Whitehead, M. 1973, Phys. Rev. Lett., $\underline{30}$, 136.

Weisfield, M. W., Kaufman, S. B., Steinberg, E. P., and Wilkins, B. D.   1975, presented at the 170th ACS National Meeting, Chicago, August 25–28.

Wilkins, B. D., Kaufman, S. B., and Steinberg, E. P.   1975, presented at the 170th ACS National Meeting, Chicago, August 25–28.

Yu, Y. W., Biswas, S., and Porile, N. T.   1975, Phys. Rev. C, $\underline{11}$, 2111.

Zebelman, A. M., Poskanzer, A. M., Bowman, J. D., Sextro, R. G., and Viola, V. E.   1975, Phys. Rev. C, $\underline{11}$, 1280.

SPALLATION REACTIONS: CALCULATIONS[*]

Hugo W. Bertini

Neutron Physics Division, Oak Ridge National
Laboratory, Oak Ridge, Tennessee, 37830

ABSTRACT.   Current methods for calculating spallation reactions over various energy ranges are described and evaluated. Recent semiempirical fits to existing data will probably yield the most accurate predictions for these reactions in general. However, if the products in question have binding energies appreciably different from their isotopic neighbors and if the cross section is ~ 30 mb or larger, then the intranuclear-cascade-evaporation approach is probably better suited.

1.   INTRODUCTION

There are two general theoretical approaches to the evaluation of spallation reactions. One is a parametric fitting of existing data, and the other is the utilization of various models to calculate the spallation cross sections.

2.   PARAMETRIC STUDY

One of the most widely known parametric studies is that of Rudstam,[1] who attempted to fit experimental data for targets that ranged from vanadium (Z = 23) to uranium and for incident protons and alpha particles with energies that varied from 60 MeV to 30 GeV. He found that of the four expressions he examined, an

---

[*] This research was funded by the U. S. Energy Research and Development Administration under contract with the Union Carbide Corporation.

Shen/Merker (eds.), Spallation Nuclear Reactions and Their Applications, 27–48. All Rights Reserved.
Copyright © 1976 by D. Reidel Publishing Company, Dordrecht-Holland.

expression of the form

$$\sigma(A,Z) = \sigma \exp[P A - R| Z - Z_p|^{3/2}] \tag{1}$$

with

$$Z_p = S A - T A^2 \tag{2}$$

best fits the data. There are five adjustable parameters in this
expression — namely, σ, P, R, S, and T. $\sigma(A,Z)$ is the spallation
cross section for producing the residual nucleus (A,Z) and Z_p is
the most probable Z, given an A. No product nearer to the target
than mass 2 is considered.

Sample results calculated from Rudstam's formula (1) are
given in Table I which contains the average ratio of the experimental to calculated quantities for the number of residual nuclides
listed. Considering the energy range involved, the results are
impressive for iron but relatively poor for bismuth. The comparison suggests that a user should utilize Rudstam's equation with
caution.

Equally well known is the more recent work of Silberberg and
Tsao,[2] who fitted data over a broader mass range with an expression of the form

$$\sigma = \sigma_0 \, f(A) \, f(E) \exp(- P \Delta A - R|Z - S A + T A^2|^{\nu})\Omega n \xi \tag{3}$$

where ΔA is the mass difference between the target and residual
nucleus. This expression is of the same general form as (1) above,
but there are 11 adjustable parameters if one includes f(A) and
f(E). Ratios of $\sigma_{calc}/\sigma_{exp}$ are generally within the range of
1 ± 0.5 using this method, which probably makes this parametric
formulation more accurate for predicting spallation yields than
any others at this time.

3.   MODELS

The models that can be used in the alternative theoretical
approach were not formulated specifically to predict spallation
data but were intended to investigate the underlying mechanisms
in nuclear reactions involving continuum-state transitions. They
are semiclassical in nature rather than quantum mechanical.

At very low energies, i.e., of the order of the Coulomb barrier, a compound nucleus evaporation model can be used, but it
will not be discussed here because of the limited energy for which
it might be useful. In the energy region from the Coulomb barrier
to about 50 MeV, various pre-equilibrium models might be employed,

Table I

Average Ratio of $\sigma_{exp}/\sigma_{calc}$ for Spallation Yields[a]

Target	Proton Irradiation	No. of Products	$\langle\sigma_{exp}/\sigma_{calc}\rangle$
Fe	130 MeV	5	2.18 ± 1.21
	208	5	1.68 ± 0.72
	297	5	0.71 ± 0.27
	340	21	0.64 ± 0.11
	396	5	0.59 ± 0.18
	500	6	0.71 ± 0.10
	590	5	0.82 ± 0.04
	600	6	1.49 ± 0.23
	730	17	0.92 ± 0.13
	24 GeV	18	0.66 ± 0.11
	25	4	1.20 ± 0.10
	29	4	1.03 ± 0.15
Bi	380 MeV	34	1.44 ± 0.23
	480	17	3.69 ± 2.01
	660	21	6.45 ± 1.82

a. G. Rudstam, Z. Naturforschg. 21a, 1027 (1966).

twc of which are discussed below.

3.1 Master equation approach

In one such approach (the HMB model), a Boltzmann-like master equation is used to govern the time evolution of the occupation numbers of single-particle energy states within excited nuclei.[3,4] The changes in the occupation numbers occur via internal two-body scatterings. Included in the equation is a term that describes the probability of emission for the particles that occupy each energy state.

3.2 Exciton models

In other models,[5-11] based primarily on the concepts of Griffin,[12] the excited nucleus is characterized by the number of excitons (i.e., excited particles within the nucleus plus holes), and the evolution toward equilibrium takes place via two-body interactions within the nucleus which successively changes the exciton number by ± 2. Again, at each state described by the exciton number, the probability that a particle will be emitted is calculated.

Evaluation. Regarding the ability of these models to calculate spallation products, they suffer from the fact that a total reaction cross section must be supplied. Furthermore, the initial configuration, be it occupation number or exciton number, must be assumed. Both objections might be circumvented[4] by initially utilizing an intranuclear-cascade calculation (described below). However, it should be kept in mind that both models were originally formulated to predict the spectra of emitted particles, rather than spallation data, for 10- to 60-MeV nonelastic nuclear reactions.

3.3 Intranuclear-cascade calculations

These calculations have been used at energies as low as 15 MeV[13] and as high as 1000 GeV.[14] However, at these energy extremes no attempt was made to calculate cross sections for spallation reactions. Rather, comparisons were made with experimental data for secondary-particle spectra and multiplicities. Results for these comparisons are encouraging. Spallation-reaction cross sections were calculated and compared to experimental data over the energy range from 25 MeV to about 3 GeV, and these comparisons will be discussed below.

Monte Carlo methods are employed in these calculations. The basic assumption is that the initial phase of the calculations (called the cascade) can be represented by a series of sequential two-body interactions. These two-body reactions are specifically calculated; i.e., the initial and final states of each two-body reaction are completely determined. Such things as the spatial point of each interaction, the type of struck particle within the nucleus (neutron or proton), the vector momentum of the struck particle, the final products of the two-body reaction (nucleons or nucleons and pions), and the vector momenta of these products are determined by the Monte Carlo selection procedures along with kinematics calculations that conserve energy and momentum.

In the HMB and exciton models, probabilities for two-body interactions are calculated and summed over all possible successive interactions. In the Monte Carlo approach, the specific life histories of each incident particle and all collision products from the two-body interactions are followed. The summing is replaced by calculating a sufficient number of life histories such that the probability distributions will have been adequately sampled and the normalization is made in terms of the total number of trials. Other important differences are that in the Monte Carlo calculation spatial locations of collision sites and trajectories are identified, whereas in the pre-equilibrium models they are not, and collision probabilities are governed by transport-type calculations, i.e., $\exp(- \Sigma x)$, rather than by available energy states or phase-space states, as they are in the pre-equilibrium models.

In the cascade calculations, after all reaction products have escaped or have been captured by the nucleus and with the nucleus in a highly excited state, the final phase of the reaction is calculated by evaporation methods.[15] It is at this stage that the pre-equilibrium models might be substituted to more realistically represent the completion of the reaction[4] since the Monte Carlo calculations supply the reaction cross sections and the "state" of the nucleus prior to subsequent decay.

Intranuclear-cascade calculations have been extensively developed at three centers: Oak Ridge National Laboratory (ORNL),[16-1] Brookhaven National Laboratory (BNL)-Columbia University,[19-23] and at the Joint Institute for Nuclear Research (JINR) in Dubna, USSR.[14,24-27] The mass-yield distributions for reaction energies of 150- and 300-MeV protons on aluminum and tantalum were compared for early versions of these models, and the results were in remarkable agreement even though there were large discrepancies in the excitation-energy distributions of the residual nuclei following the cascade reaction.[28]

The three calculational versions are similar in that the diffuse edge of the nucleus, the exclusion effects, the motion of the
bound nucleons within the nucleus, the spatial variations of the
single-particle nuclear potential inside the nucleus, and pion
production and pion-nucleon interactions are all taken into account.
They differ in that reflection and refraction effects and a velocity-
dependent nuclear potential are included only in the BNL-Columbia
version; the ORNL version is the only one that does not prohibit
two or more interactions from taking place within the same spatial
"volume" (a volume equivalent to that occupied by a bound nucleon)
inside the nucleus, although efforts to include this prohibition
are underway at ORNL; and the JINR version is the only one that
permits two or more cascade particles to interact simultaneously
with a single nucleon inside the nucleus. Also, the JINR version
is the only one of the three that assigns specific locations for
the nucleons of the nucleus, with the locations selected from appropriate charge distributions each time a new incident particle
is to be traced.[27] Recent versions of the BNL-Columbia model[22,23]
have included isobar[29] (excited nucleon states) interactions with
nucleons but have not included reflection and refraction effects
or a velocity-dependent potential.

For the calculational data that are available, the ORNL version and the BNL-Columbia version that includes reflection and refraction, a velocity-dependent potential, and nucleon correlations
(i.e., preventing cascade particles from interacting twice or more
in the same nucleon "volume") appear to give the best agreement
with each other and with experimental data in general at energies
~ 200 MeV. As already mentioned, these versions also agree with
results from JINR with regard to mass yields.[28] Similarly, results from the BNL-Columbia version that includes isobar interactions, the JINR version, and those from ORNL appear to agree at
energies ~ 1 GeV.

4.   COMPARISON WITH EXPERIMENTAL DATA

Since differences among results from the three intranuclear-
cascade models do not appear significant, comparisons with experimental data for spallation yields will be illustrated for only the
ORNL results. In order to ascertain systematic differences, if
any, comparisons for the energy range of ~ 25 MeV to ~ 400 MeV for
incident protons on targets with increasing mass are illustrated
first. Results for simple reactions (p,pn) and more complicated
reactions (p,xpyn) in this energy range for each target are shown.
This is followed by comparisons for incident protons with energies
from ~ 500 MeV to ~ 3 GeV on a few targets, and these data, in
turn, are followed by comparisons for incident pions from 25 MeV
to 1.6 GeV.

Figures 1-8 illustrate the comparison for incident protons over the energy range from 25 to 400 MeV.[30] Except for the $^{12}C(p,3p3n)^7Be$ reaction (Fig. 2), the agreement is moderate to good. The ORNL version has been consistently poor in predicting cross sections for producing beryllium in spallation reactions, and the reasons for this are not understood.

Figures 9 and 10 and Tables II and III illustrate comparisons for incident protons at energies ranging from 500 MeV to 2.9 GeV,[18] and the agreement varies from poor to excellent. The latest BNL-Columbia version[23] gives better agreement for the production of ^{24}Na and ^{15}O from 0.6 and 1 GeV protons on ^{27}Al (Fig. 9).

Figures 11 and 12 and Table IV illustrate comparisons for incident π^- with energies ranging from 25 MeV to 2.6 GeV, with reasonable agreement in general. In the energy range shown in Fig. 11, the BNL-Columbia calculation gives better agreement with experimental data for the (π^-,π^-n) reaction on carbon. Comparison of the ORNL results with recently measured $^{12}C(\pi^\pm,\pi N)^{11}C$ reactions in the same energy range[31] (not illustrated) indicate that the model overestimates the $^{12}C(\pi^-,\pi^-n)^{11}C$ cross section by about 30% at the peak value and underestimates the $^{12}C(\pi^+,\pi N)^{11}C$ cross section by about 20%, which leads to a calculated ratio for $\sigma(\pi^-)/\sigma(\pi^+)$ of 2.6 compared to a measured ratio of 1.55. Effects of nucleon charge exchange, offered as an explanation of this long-standing discrepancy,[32] are, in fact, included with little approximation in the intranuclear-cascade calculations. Therefore, these effects should not be major contributors to the discrepancy.

The discrepancy between the predictions and the experimental data at higher energies for the $^{12}C(\pi^-,x)^{11}C$ reaction (Fig. 12) can be attributed to the fact that multiple pion production in the two-body reactions is not included in the calculations. Such reactions would increase the complexity of the cascade phase, making it less likely that ^{11}C is produced as a final product from a ^{12}C target.

Table V illustrates comparisons from another cascade model (RENO), which is still under development.[33] This model employs a discrete nucleon representation for the target nucleus, as does the JINR version,[14,24-28] but it can also do the same for complex incident particles. Typical results are shown in Table V, where there is reasonable agreement between the predictions from the model and experimental data.

Finally, results from a preliminary version of a heavy-ion reaction model[34] are shown in Table VI. In this model, the individual nucleons of complex incident nuclei are treated independently as they generate cascade reactions within the target nucleus. The total reaction is considered to be the sum of the individual

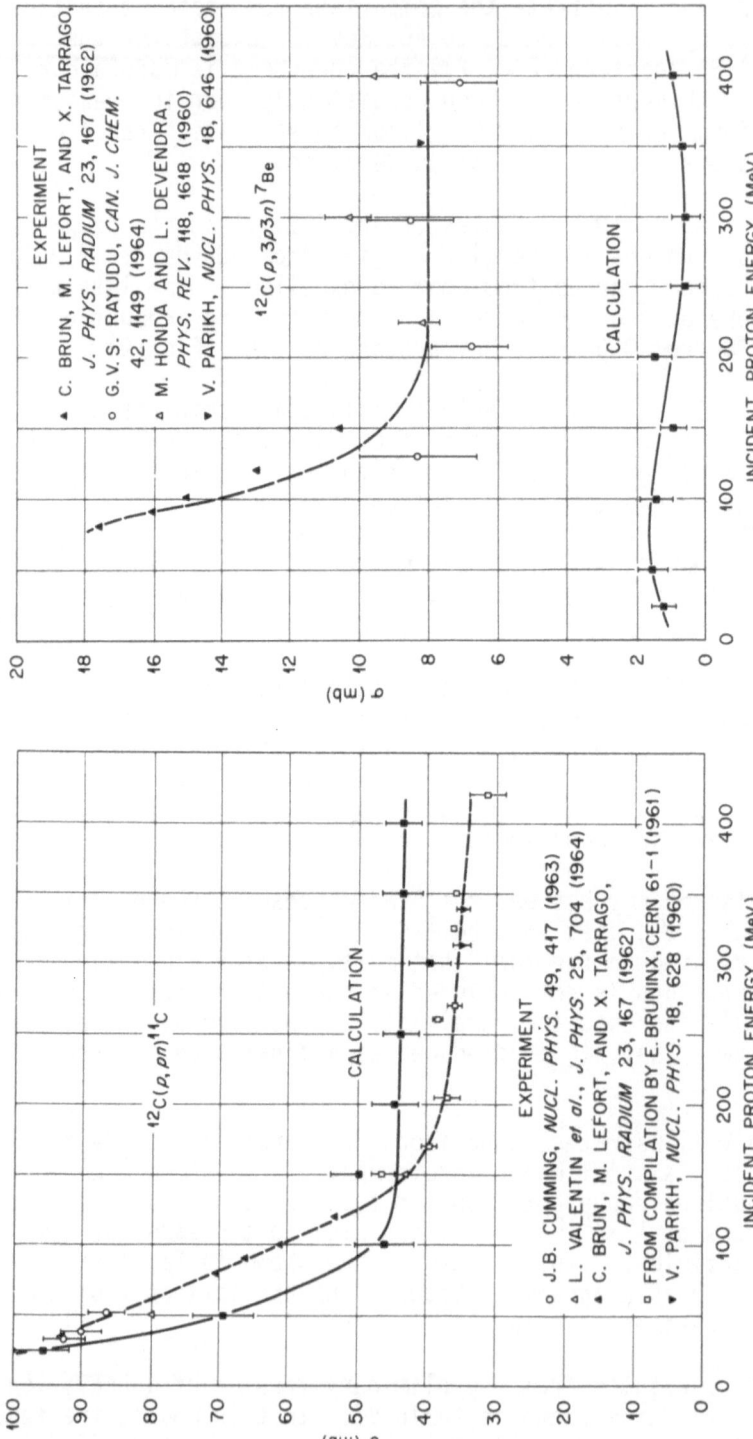

Fig. 1. Cross section for the ^{12}C(p,pn)^{11}C reaction vs incident proton energy.

Fig. 2. Cross section for the ^{12}C(p,3p3n)^{7}Be reaction vs incident proton energy.

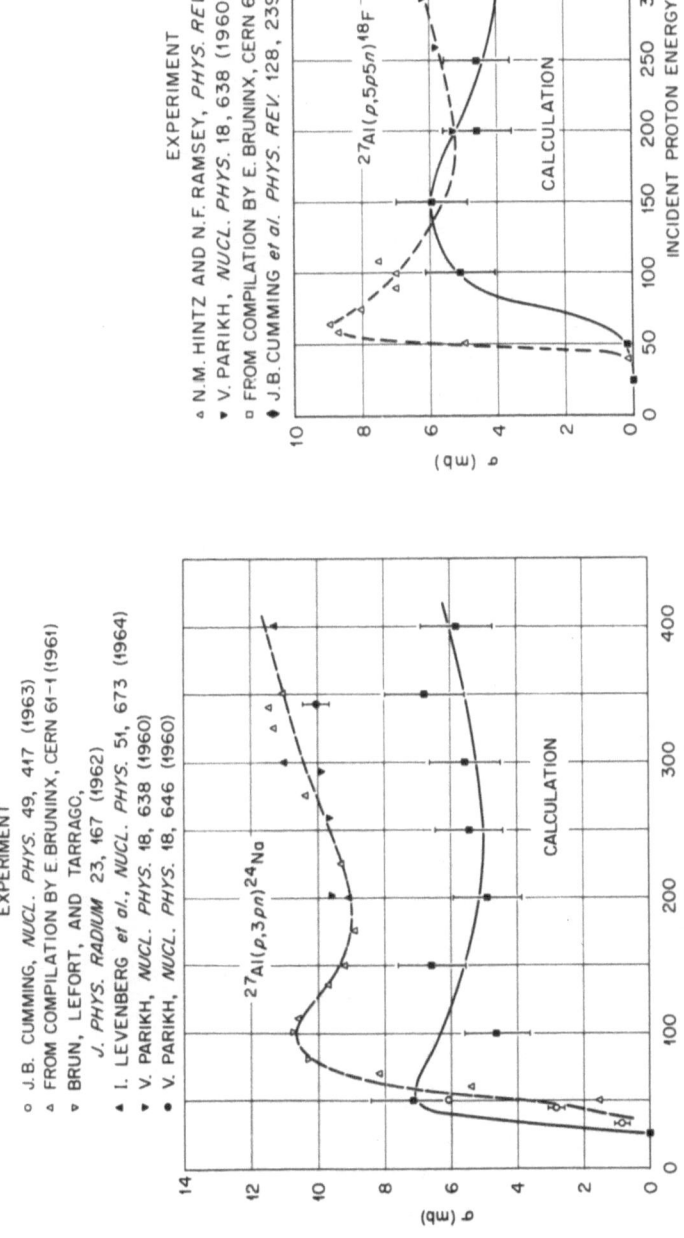

Fig. 4. Cross section for the $^{27}Al(p,5p5n)^{18}F$ reaction vs incident proton energy.

Fig. 3. Cross section for the $^{27}Al(p,3pn)^{24}Na$ reaction vs incident proton energy.

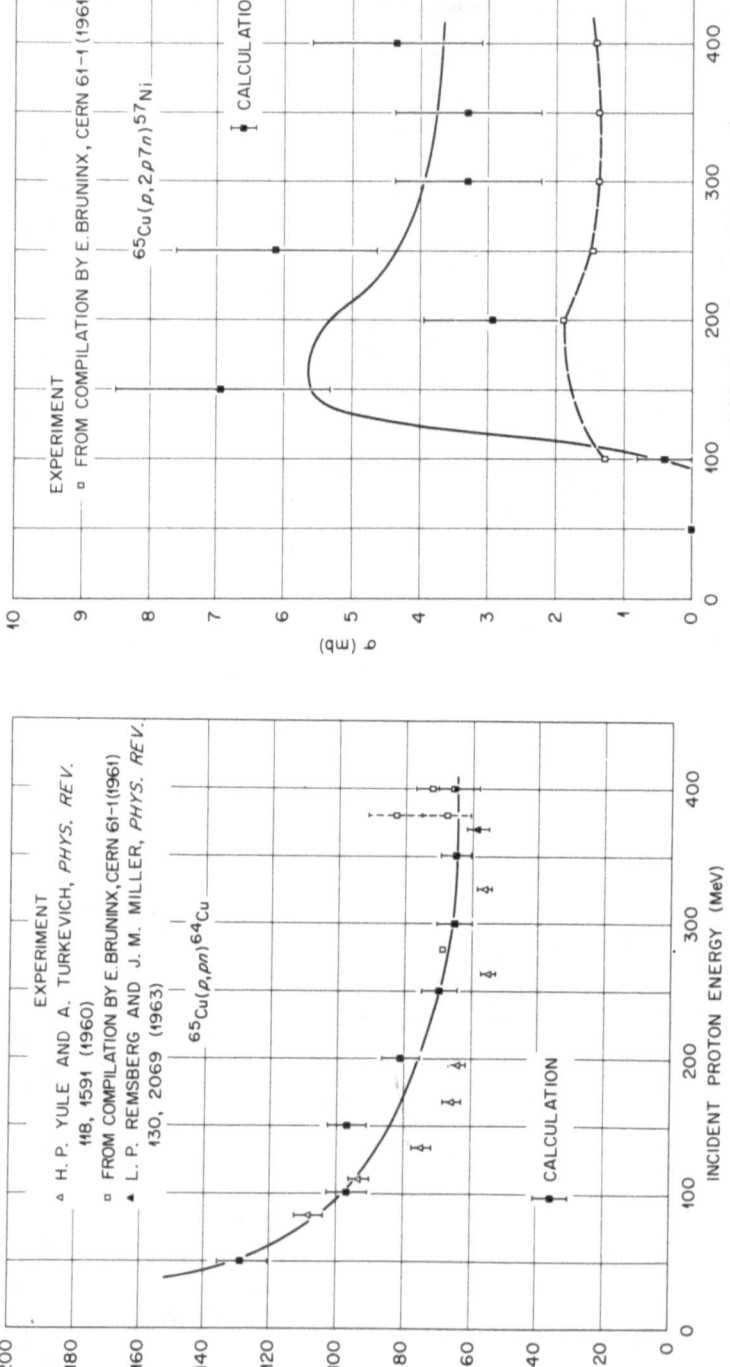

Fig. 6. Cross section for the $^{65}Cu(p,2p7n)^{57}Ni$ reaction vs incident proton energy.

Fig. 5. Cross section for the $^{65}Cu(p,pn)^{64}Cu$ reaction vs incident proton energy.

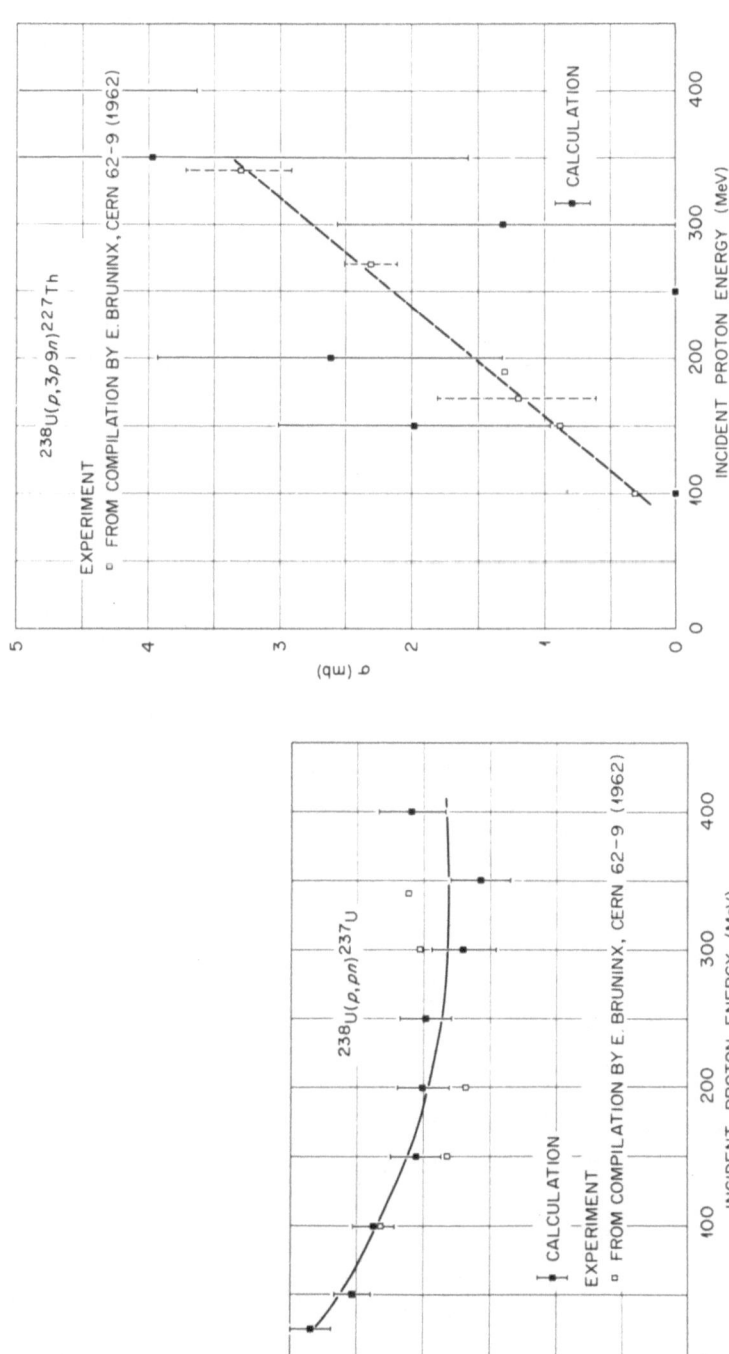

Fig. 8. Cross section for the $^{238}U(p,3p9n)^{227}Th$ reaction vs incident proton energy.

Fig. 7. Cross section for the $^{238}U(p,pn)^{237}U$ reaction vs incident proton energy.

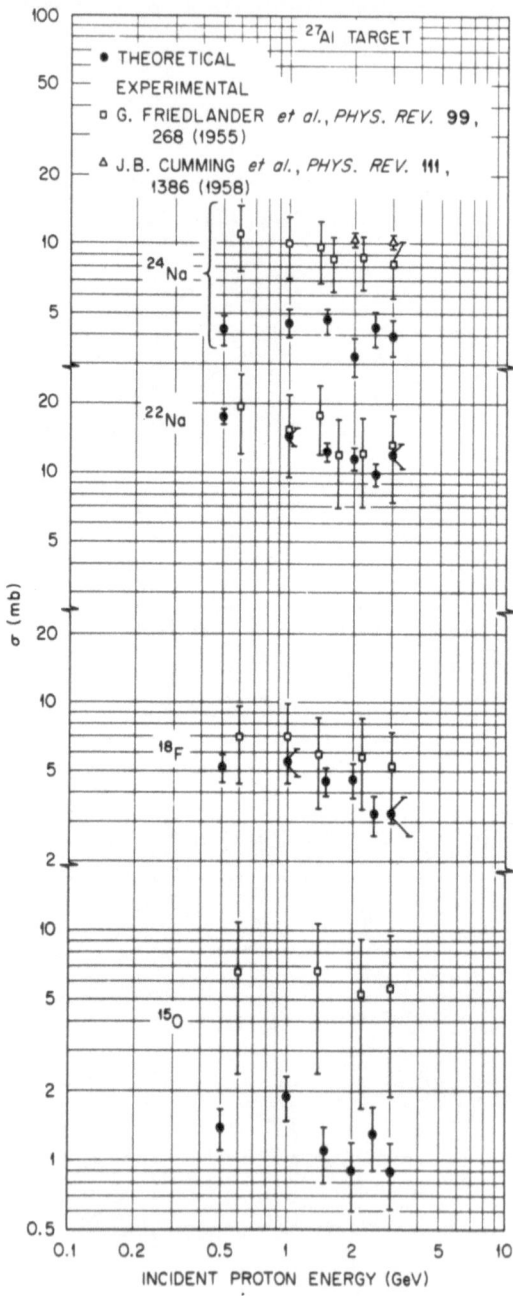

Fig. 9. Spallation cross sections for
protons on aluminum.

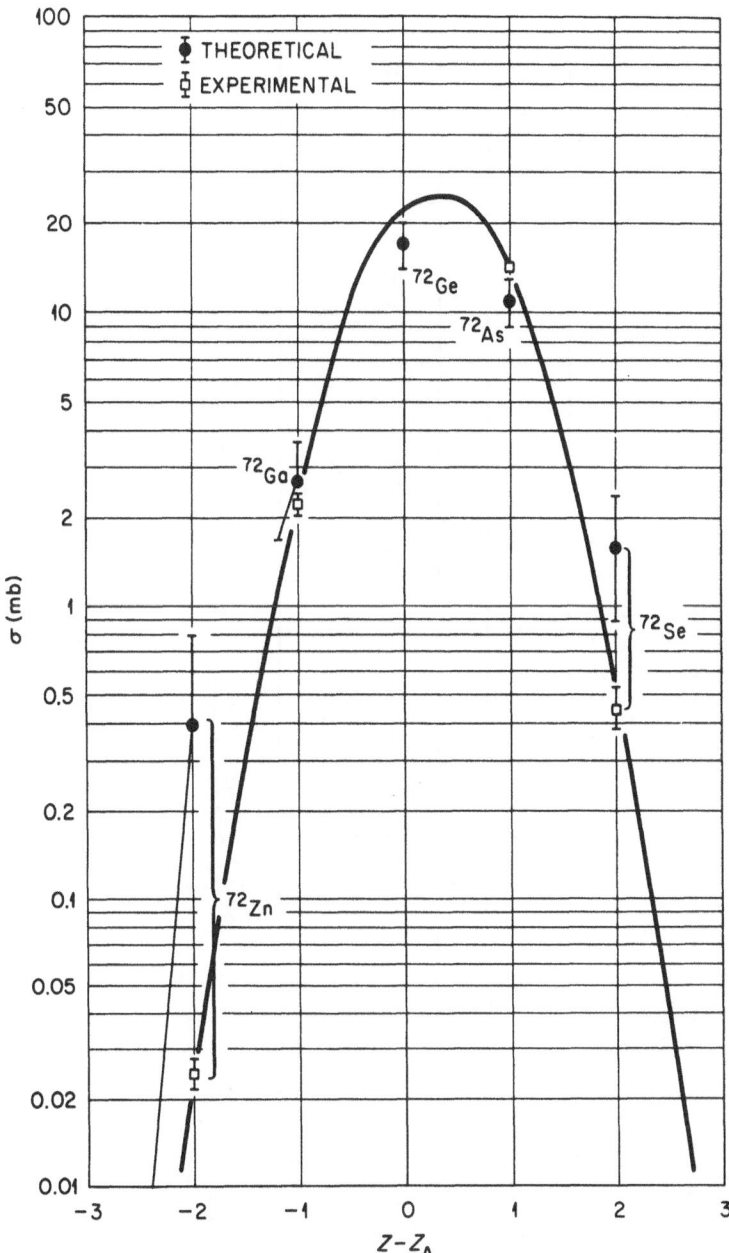

Fig. 10. Spallation cross sections for the isotopes of mass 72 from 2.9-GeV protons on arsenic. Experimental data are from S. Kaufman, Phys. Rev. 126, 1189 (1962). The solid line, from Kaufman, is the Gaussian curve that best fits the experimental points. He takes $Z_A = 32$.

Table II

Experimental[a] and Theoretical Cross Sections for the Formation
of Iodine Isotopes From 720- and 2000-MeV Protons on ^{127}I

Cross Sections
(mb)

Proton Energy (MeV)	$^{127}I(p,pn)^{126}I$		$^{127}I(p,p2n)^{125}I$		$^{127}I(p,p3n)^{124}I$		$^{127}I(p,p6n)^{121}I$	
	Exp.	Theor.	Exp.	Theor.	Exp.	Theor.	Exp.	Theor.
720	55 ± 14	78 ± 7	15 ± 4	29 ± 4	19 ± 5	20 ± 4	13 ± 3	10 ± 3
2000	59 ± 15	61 ± 6	20 ± 5	26 ± 4	18 ± 5	22 ± 4	11 ± 3	7 ± 2

a. I. Ladenbauer and L. Winsberg, Phys. Rev. 119: 1368 (1960).

Table III

Experimental[a] and Theoretical Cross Sections for the Formation of Tellurium Isotopes From 720- and 2000-MeV Protons on ^{127}I

Proton Energy (MeV)	$^{127}I(p,2p9n)^{117}Te$ + $^{127}I(p,2p10n)^{116}Te$		$^{127}I(p,2p8n)^{118}Te$		$^{127}I(p,p\pi^+)^{127}Te$	
	Exp.	Theor.	Exp.	Theor.	Exp.[b]	Theor.
720	23.6 ± 4.0	43 ± 5	31.1 ± 7.8	18 ± 3	< 3.2 ± 1.1	0.7 ± 0.7
2000	7.0 ± 1.5	12 ± 3	18.1 ± 4.5	12 ± 3	< 3.2 ± 1.1	1.3 ± 0.9

Cross Sections (mb)

a. I. Ladenbauer and L. Winsberg, Phys. Rev. 119: 1368 (1960).

b. These values represent the sum of the stable and metastable states of ^{127}Te.

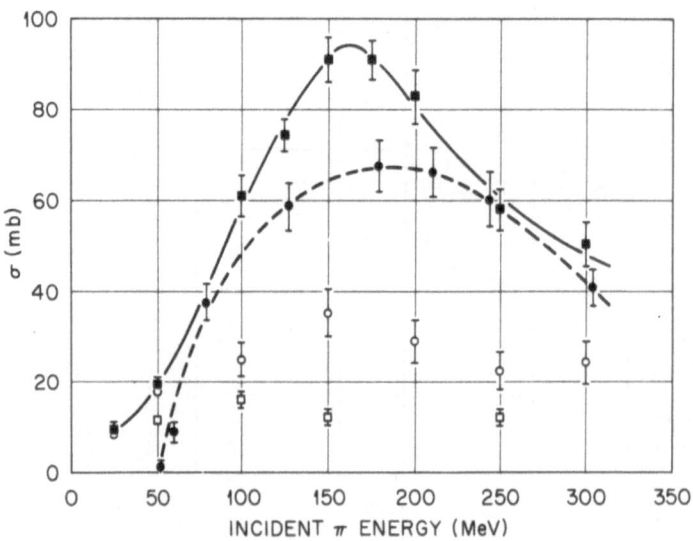

Fig. 11. Cross section for the $^{12}C(\pi^-,\pi^-n)^{11}C$ reaction vs incident pion energy.

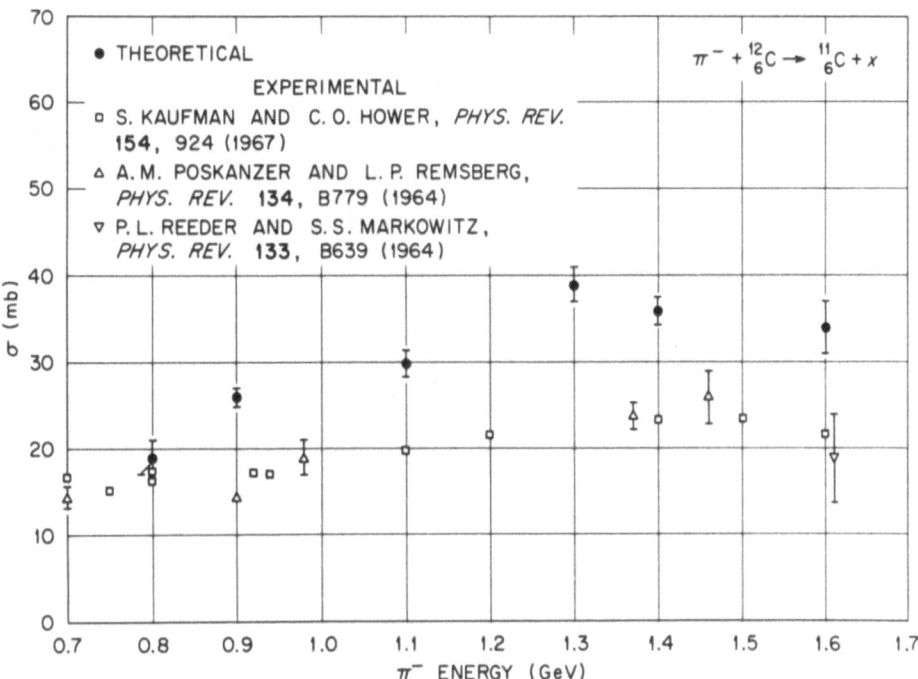

Fig. 12. Cross sections for the production of ^{11}C from incident π^- mesons. The experimental error in the data of Kaufman and Hower is about the size of the symbols.

Table IV

Experimental and Theoretical Radiochemical Cross
Sections for π^- on Carbon and Argon

Cross Sections
(mb)

π^- Energy (MeV)	$^{12}C(\pi^-,\pi^-n)^{11}C$		$^{40}Ar(\pi^-,\pi^-p)^{39}Cl$		$^{40}Ar(\pi^-,\pi^-pn)^{38}Cl$	
	Exp.[a]	Theor.	Exp.[b]	Theor.	Exp.[b]	Theor.
600						
800	16.1 ± 1.0	19 ± 2	24.5 ± 1.2	47 ± 3	14.3 ± 0.9	20 ± 2
1100						
1600	20.5 ± 0.8	34 ± 3	20.8 ± 1.1	33 ± 3	13.0 ± 1.6	16 ± 2

a. S. Kaufman and C. O. Hower, Phys. Rev. 154: 924 (1967).

b. C. O. Hower and S. Kaufman, Phys. Rev. 144: 920 (1966).

Table V

Comparisons of Predictions from the RENO-II Model with
Experimental Data for 150-MeV Protons on ^{16}O

Residual Nuclide	Cross Section (mb)	
	Experimental Data[a]	RENO-II
$^{10}B + ^{10}C$	10	16 ± 2
$^{11}B + ^{11}C$	25	30 ± 3
^{12}C	30	36 ± 3
$^{13}C + ^{13}N$	25	17 ± 2
^{14}O	33	29 ± 3
$^{15}N + ^{15}O$	52	51 ± 4

a. R. Silberberg and C. H. Tsao, Astrophys. J. 25, Suppl. 220, 315 (1973).

Table VI

Cross Section (mb) for the Production of Various Fragments with Velocities Within ± 2% (for A Values of 12 and 13) and ± 4% (for A ≤ 11) of the Incident ^{14}N Velocity and Emitted into the Angular Interval of 0-4 mrad from the Reactions of ^{14}N at 2.1 GeV/Nucleon on ^{12}C

Fragment	Theoretical Results for the Indicated Nuclear Configurations			Experimental Data[a]
	Standard	Smaller	Smallest	
$^{7}_{3}$Li	0	0.35 ± 0.18	0.30 ± 0.15	9.1 ± 3
$^{7}_{4}$Be	0	0.09 ± 0.09	0.07 ± 0.07	8.0 ± 1.0
$^{9}_{4}$Be	0	0.09 ± 0.09	0.30 ± 0.15	3.0 ± 0.3
$^{8}_{5}$B	0	0	0	5.1 ± 0.6
$^{10}_{5}$B	13 ± 3	12 ± 1	9.8 ± 0.9	14.4 ± 2.9
$^{11}_{5}$B	8.5 ± 2.1	6.5 ± 0.8	4.9 ± 0.6	10.7 ± 1.3
$^{12}_{5}$B	0.5 ± 0.5	2.2 ± 0.4	1.3 ± 0.3	1.9 ± 0.4
$^{9}_{6}$C	0	0.18 ± 0.12	0	0.14 ± 0.02
$^{10}_{6}$C	0	0.35 ± 0.18	0.37 ± 0.17	0.96 ± 0.12
$^{11}_{6}$C	7.4 ± 2.0	5.4 ± 0.7	4.0 ± 0.5	11.3 ± 1.1
$^{12}_{6}$C	68 ± 6	58 ± 2	54 ± 2	46 ± 9
$^{13}_{6}$C	36 ± 4	21 ± 1	17 ± 1	9.2 ± 1.8
$^{12}_{7}$N	0.5 ± 0.5	0.79 ± 0.26	0.22 ± 0.13	0.63 ± 0.13
$^{13}_{7}$N	17 ± 3	11 ± 1	10 ± 1	7.7 ± 1.5
Sum	151 ± 9	118 ± 3	102 ± 3	128.13 ± 10.4
Total Re- action Cross Section	1112	966	918	
% $\frac{Sum}{Total}$	14%	12%	11%	

a. H. H. Heckman *et al.*, "Fragmentation of ^{14}N Nuclei at 29 GeV: Inclusive Isotope Spectra at 0°," Phys. Rev. Letters 28, 926 (1972).

contributions. The data in the middle column for the theoretical results, labeled "Smaller," are those that should be compared with the experimental results. The agreement is fair.

5.   SUMMARY

If one is interested in the accurate predictions of spallation yields, the parametric fitting procedure of Silberberg and Tsao[2] is probably the most appropriate. One can have greater confidence in this method if the reaction energy and target are within the limits of the experimental data that were used to establish the parameters and if the binding energy of the spallation nucleus in question differs little from that of its neighbors. When there are significant differences in binding energies among the neighboring products, any of the intranuclear-cascade models would probably be more reliable, particularly if these cross sections represent at least 10% of the total reaction cross section.[30]

The suggestion that the state of the excited residual nucleus, as determined by intranuclear-cascade calculations, be used as the starting point for pre-equilibrium calculations[4] is surely worthwhile, but it has never been fully explored.

The validity of other approaches[33,34] for the calculation of spallation yields from reactions between complex nuclei cannot be evaluated until these programs are completed and extensive comparisons with experimental data are made. However, preliminary results are encouraging.

REFERENCES

1.   G. Rudstam, Z. Naturforschg. 21a, 1027 (1966).
2.   R. Silberberg and C. H. Tsao, Astrophys. J. 25, Suppl. 220, 315 (1973).
3.   G. D. Harp, J. M. Miller, and B. J. Berne, Phys. Rev. 165, 1166 (1968).
4.   G. D. Harp and J. M. Miller, Phys. Rev. C3, 1847 (1971).
5.   Marshall Blann, Phys. Rev. Letters 21, 1357 (1968).
6.   C. K. Cline and M. Blann, Nucl. Phys. A172, 225 (1971).
7.   C. K. Cline, Nucl. Phys. A174, 73 (1971).
8.   Marshall Blann, Phys. Rev. Letters 27, 337 (1971).
9.   G. M. Braga-Marcazzan *et al.*, Phys. Rev. C6, 1398 (1972).
10. Marshall Blann, Phys. Rev. Letters 28, 757 (1972).
11. E. Gadioli, E. Gadioli-Erba, and P. G. Sona, Nucl. Phys. A217, 589 (1973).
12. J. J. Griffin, Phys. Rev. Letters 17, 478 (1966).
13. R. G. Alsmiller, Jr. and O. W. Hermann, Nucl. Sci. Eng. 40, 254 (1970).

14. V. S. Barashenkov, A. S. Iljinov, and V. D. Toneev, "In-elastic Interactions of Particles and Nuclei at High and Superhigh Energies," Communication of the Joint Institute for Nuclear Research, E2-5282, Dubna, USSR (1970).

15. I. Dostrovsky, Z. Fraenkel, and G. Friedlander, Phys. Rev. 116, 683 (1959).

16. Hugo W. Bertini, Phys. Rev. 131, 1801 (1963); with errata 138, AB2 (1965).

17. Hugo W. Bertini, Phys. Rev. 188, 1711 (1969).

18. Hugo W. Bertini, Phys. Rev. C6, 631 (1972).

19. K. Chen *et al.*, Phys. Rev. 166, 949 (1968).

20. K. Chen, G. Friedlander, and J. M. Miller, Phys. Rev. 176, 1208 (1968).

21. K. Chen *et al.*, Phys. Rev. C4, 2234 (1971).

22. G. D. Harp *et al.*, Phys. Rev. C8, 581 (1973).

23. G. D. Harp, Phys. Rev. C10, 2387 (1974).

24. V. S. Barashenkov, K. K. Gudima, and V. D. Toneev, Acta Physica Polon. 36, 415 (1969).

25. V. S. Barashenkov, K. K. Gudima, and V. D. Toneev, "Computing Process for Intranuclear Cascades," JINR preprint P2-4065 (1968).

26. V. S. Barashenkov, A. S. Iljinov, and V. D. Toneev, "Further Developments of the Intranuclear Cascade Model," JINR preprint P2-5280 (1970).

27. V. S. Barashenkov *et al.*, "Calculation of Nuclear Boundary Diffusion by the Model of Intranuclear Cascades," JINR preprint P2-6503 (1972).

28. V. S. Barashenkov *et al.*, Nucl. Phys. A187, 531 (1972).

29. R. M. Sternheimer and S. J. Lindenbaum, Phys. Rev. 123, 333 (1961).

30. Hugo W. Bertini, Phys. Rev. 171, 1261 (1968).

31. B. J. Dropesky *et al.*, Phys. Rev. Letters 34, 821 (1975).

32. Morton M. Sternheim and Richard R. Silbar, Phys. Rev. Letters 34, 824 (1975).

33. William F. Schmitt, "A Monte Carlo Calculation of Spallation Reactions of Astrophysical Interest in the M Nuclei," a dissertation, Graduate School of Arts and Sciences, University of Pennsylvania, Philadelphia, PA (1970); W. F. Schmitt *et al.*, "A Model for Astrophysical Spallation Reactions," 13th International Cosmic Ray Conference, August 17-30, 1973, Denver, CO, Conf. 730804; C. L. Ayres *et al.*, "Calculation of Astrophysical Spallation Reactions Using the Reno Model," ibid.

34. H. W. Bertini *et al.*, "HIC-1: A First Approach to the Calculation of Heavy-Ion Reactions at Energies ≥ 50 MeV/Nucleon," ORNL-TM-4134, Oak Ridge National Laboratory (1974); H. W. Bertini, T. A. Gabriel, and R. T. Santoro, Phys. Rev. C9, 522 (1974).

SEMIEMPIRICAL CROSS SECTIONS, AND APPLICATIONS TO NUCLEAR INTERACTIONS OF COSMIC RAYS

R. Silberberg, C. H. Tsao, and M. M. Shapiro

Laboratory for Cosmic Ray Physics, Naval Research Laboratory, Washington, D.C. 20375

ABSTRACT. We have formulated semiempirical calculations of partial cross sections (papers ST in the references) for calculating the many yields, as yet unmeasured, of high-energy reactions. Here we present parameters that are updated on the basis of recent experimental data. These calculations have been employed in the analysis of cosmic-ray data when measured values are unavailable. (About half of the cosmic rays heavier than helium have undergone spallation since their acceleration.) The calculations have yielded: (a) the elemental composition of cosmic rays at the sources, (b) the expected isotopic composition of cosmic rays at the top of the atmosphere, (c) the mean path length of material traversed by cosmic rays, and the distribution function of path lengths, and (d) predicted abundances of long-lived radioactive secondary cosmic-ray nuclides for the determination of confinement time in the galaxy. The current status of these calculations is presented, using the experimental data on cross sections and on the composition of cosmic rays reported up to September, 1975.

1. INTRODUCTION

High-energy nuclear reaction cross sections are essential for interpreting observations of cosmic-ray physics and astrophysics. They are also vital in planetary physics and in lunar and meteoritic studies. However, most of the cross sections needed for calculating the transformations of cosmic rays in collisions with the interstellar gas have not yet been measured. When measured cross sections are available, they should be used, of course. The unmeasured partial cross sections can be calculated by either

Shen/Merker (eds.), Spallation Nuclear Reactions and Their Applications, 49–81. All Rights Reserved.
Copyright © 1976 by D. Reidel Publishing Company, Dordrecht-Holland.

Monte Carlo or semiempirical techniques. Which of the two should
be employed depends on the nature of the reactions and the energy -
we shall return subsequently to this problem. However, Monte
Carlo calculations are exceedingly time consuming even on large
computers and have been carried out only for a small set of target-
product-energy combinations. Also, Monte Carlo calculations of
partial cross sections have not yet been extended with good precision to energies above 1 GeV, owing to complications introduced
by the multiple production of pions and strange particles. Hence,
experimental and semiempirical methods have generally been employed
in the study of interactions of high-energy cosmic ray nuclei.

Among the cosmic ray nuclei heavier than helium, nearly half
have suffered nuclear spallation reactions since their acceleration. Hence, to unravel the details of propagation, isotopic
composition, sources and mode of acceleration, a large number of
spallation cross sections must be known. By combining (a) the
experimental data on the isotopic composition of cosmic rays at
the top of the atmosphere and (b) diffusion equation calculations
employing spallation cross sections, one may find answers to
several questions of astrophysical interest: Are cosmic rays
generated near supernovae by acceleration of nuclei newly formed
by nucleosynthesis, or are they derived from old interstellar gas?
Does the source composition of cosmic rays require processes of
nucleosynthesis different from those that gave rise to the material
of the solar system? How long are cosmic rays confined by the
magnetic fields in the galaxy? How much time has passed since
the nucleosynthesis of the cosmic ray nuclides? How much material
have cosmic rays passed through since their acceleration; is this
material mainly in the interstellar gas, or also in cosmic-ray
source regions? Are the cosmic rays confined principally to the
galactic disc, or are there magnetic structures in the low-density
galactic halo that confine the particles? To what extent are
cosmic rays decelerated by the out-flowing solar wind? The ten-
tative answers that are available to some of these questions will
be discussed below, and crucial experiments for resolving the
remaining questions are outlined.

2. CALCULATION OF PARTIAL CROSS SECTIONS

2.1 Semiempirical calculations

Rudstam observed that there are systematic regularities among the
relative yields of nuclear reactions that depend on the neutron-
to-proton ratio of the product nuclides, and on the target-product
mass difference. He constructed an equation with several empirical parameters (Rudstam, 1955, 1956, 1966) for calculating partial
cross sections at energies $E \gtrsim 100$ MeV. These parameters are

applicable to proton interactions with nuclei heavier than calcium, except when the target-product mass difference ΔA is small
or large, i.e., it is not applicable for $\Delta A \lesssim 5$ and $\Delta A \gtrsim 40$ (the
limiting values depend on the target mass). An extrapolation of
Rudstam's relation into regions where it is inapplicable can
result in huge errors: The yield of ^7Be from ^{56}Fe at 150 MeV is
underestimated by a factor of about 10^6.

We (Silberberg and Tsao, 1973 a and b, hereafter referred
to as papers ST-1 and ST-II) have constructed a semiempirical
equation resembling Rudstam's (1966) with additional parameters,
and we have defined regions of target and product mass intervals
where these parameters apply. The basic equation for calculating
the partial cross sections is:

$$\sigma = \sigma_0 \, f(A) f(E) \, e^{-P\Delta A} \, \exp(- R| Z - SA + TA^2|^{\upsilon}) \, \Omega \, \eta \, \xi . \quad (1)$$

It is applicable for calculating cross sections of targets having
mass nos. in the range $9 \le A_t \le 209$ and products with $6 \le A \le 200$,
except for peripheral interactions with small values of ΔA (i.e.,
of $A_t - A$). For the latter reactions, a different equation was
constructed. (A different equation was devised also for target
elements as heavy as Th and U.)

In Figure 1, the experimental cross sections of proton-
induced reactions at 150 MeV in iron are compared with those
based on our equation and parameters, as well as with those based
on Rudstam's equation. For an iron target, Rudstam considered
his relation to be applicable only to products with $A \gtrsim 22$. In
fact, very large deviations would arise if his relation were
extrapolated beyond the applicable region, as illustrated by the
dashed curves for Be and F.

In equation (1), σ_0 is a normalization factor. The factors
$f(A)$ and $f(E)$ apply only to products from heavy targets (with
atomic number $Z_t > 30$, when ΔA is large, as in the case of fission
and evaporation of light product nuclei. The factor $\exp(-P\Delta A)$
reflects the diminution in cross sections as the difference of
target and product mass increases. The next exponential factor,
due to Rudstam (1966), describes the distribution of the cross
sections of various product isotopes with mass number A, that
belong to an element of atomic number Z; Figure 1 illustrates
such distribution curves (R, S, and T are discussed in ST-1,
pp 317-320). The parameter Ω is related to the nuclear structure
and number of particle-stable levels of a product nuclide. The
factor η depends on the pairing of protons and neutrons in the
product nucleus; it is larger for even-even nuclei. The parameter
ξ is introduced to represent the enhancement of light evaporation
products.

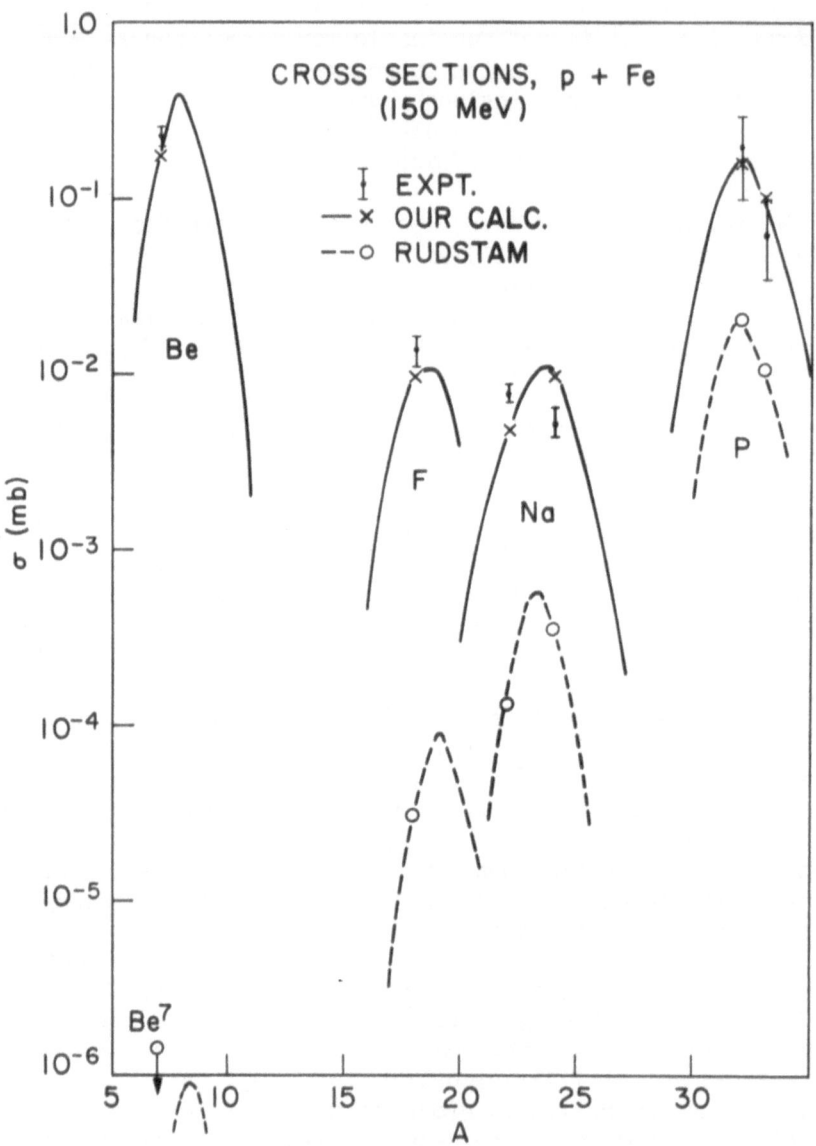

Figure 1. The partial cross sections of p-Fe interactions at
150 MeV. The experimental values are compared with those based
on our semiempirical formulas (shown by the solid curve) and
with those based on the Rudstam relation (the dashed curved).
For A ⩽ 22, Rudstam's equation is not applicable. The dashed
curves for Be and F only illustrate the danger of extrapolating
his equation into inapplicable regions.

An empirical equation for the total inelastic cross section of p-A_t reactions, applicable to $A_t \geq 6$, above 100 MeV/u, in units of mb, is:

$$\sigma_{\text{inel.}} = 10 \, \pi \, [1.31 \, A_t^{\,1/3}]^2 \, (1 - 0.47 \, A_t^{\,-0.4}) \tag{2}$$

2.2 Recent modifications of semiempirical calculations

The numerical values and conditions of applicability of the parameters of equation (1) are discussed in paper ST. Here we shall only present the newer modifications that we have introduced on the basis of recent experimental data. Since nearly all cosmic-ray nuclei (except for about one in 10^6) have atomic numbers $Z_t \leq 28$, we restrict the present discussion of new modifications to this set of elements.

The new experimental data on partial cross sections that have permitted improvements in our equations are from Lindstrom et al. (1975), who measured all the cross sections of ^{12}C and ^{16}O, Yiou et al. (1973) and Yiou and Raisbeck, (1972) who measured the yields of isotopes of Be and Li from various light elements, and Radin et al. (1974) who measured the yields of ^{7}Be from several targets. Furthermore, the measurements of the spallation of iron by Perron (1975), Raisbeck et al. (1975), and Lagarde-Simonoff et al. (1975) have permitted additional improvements.

The modifications are the following:

New values of the nuclear structure factor Ω, shown in Table 1, that replace certain values in Table 2 of paper ST-I.

For targets with $Z_t \leq 7$, the cross sections of (p, 3p) reactions are raised by 1.5.

TABLE 1

New Values for the Nuclear Structure Factor Ω of Equation (1)[*]

Product	^{7}Li	^{7}Be	^{9}Be	^{10}B	^{11}B	K	Sc
Ω	1.8	0.95	0.65	0.7	1.7	0.7	0.7

[*] These are revisions of (and, in the case of ^{7}Be, an addition to) some of the values shown in Table 2 of ST-I. Other values of that table remain unchanged.

In the case of breakup of ^{12}C, ^{14}N and ^{16}O, the cross sections $(p,p\alpha) + (p,3p2n)$ and $(p,pd) + (p,2pn)$ are enhanced by a factor of 1.8; the yields of $(p,2p)$ reactions are enhanced by 1.5, unless the product nuclide is of the type $(x \alpha + n)$, (e.g., ^{13}C, where x = 3), in which case, due to the probability of neutron evaporation, the rate is reduced to 0.5 instead.

Only for product nuclei with one or a few particle stable energy levels (i.e., for ^{13}N and ^{19}Ne), are the cross sections $\sigma_{p,pn}(E_0)$ and $\sigma_{p,2p}(E_0)$ in Eqs. 24 and 26 of paper ST-II multiplied by Ω, and of the light products in Table 1A of paper ST-I, only ^9Be is now multiplied by Ω/η.

For light nuclei ($Z_t \leqslant 5$) with a low p/n ratio ($N_p/N_n < 1$), $\sigma_{p,2p}(E_0)$ is multiplied by $(N_p-2)/(N_n-2)$ where N_p and N_n are respectively the numbers of protons and neutrons in the target nucleus. The constant 2 is based on assuming a ^4He core, which does not participate in peripheral interactions.

A minus sign was omitted in equation (23) of paper ST-II. It should read:

$$Y(A_t, Z_t) = 2 - \exp [-d (A_t - \overline{A}_t)/Z_t].$$

For $35 \leq A_t \leq 70$, not only $(p,3p)$ but also $(p,4p)$ and $(p, 5p)$ $(p,5p)$ are calculated from equation (21) of ST-II, with

$$\sigma(E_0) = \sigma_s(E_0), \text{ and } H(E) = \left\{ 1 - \exp [(-\frac{E}{35})^4] \right\} (\frac{E}{200}) \leq 1. \text{ (For}$$

σ_s, see Eq. (2a) of ST-II.)

The enhancement factors ξ of Table 3, paper ST-1 for light evaporation products from targets with $34 \leq A_t \leq 63$ are now: $3 [1 + 0.02 (A_t - 34)]$ for ^6He, $1 + 0.02 (A_t - 34)$ for ^6Li, and $1 + 0.01 (A_t - 34)$ for ^7Li and ^7Be.

For the production of boron, except by (p,pn) and $(p,2p)$ reactions, the geometric mean of calculations based on Table 1A combined with 1B, 1C and 1D of paper ST-I is to be used for Z_t = 6 to 16, 17 to 20, and 21 to 28, respectively. (Previously the type of mean to be used was not specified.)

In the case of spallation of ^{56}Fe, the products ^{49}V, ^{50}V and ^{51}V are multiplied by 0.8, 0.7 and 0.5 respectively, ^{54}Mn and ^{55}Mn by 1.5 and 1.1, ^{53}Fe by 2, and Cl by 0.9.

2.3. Extension of semiempirical calculations to nucleus-nucleus interactions

The cross sections of nucleus-nucleus reactions are also of

considerable importance in analyzing cosmic-ray data. The
experiments are carried out on balloon flights or in spacecraft
with material intervening above the detector. Collisions of
nuclei with helium, too, are of interest, since an appreciable
fraction (~ 20%) of cosmic-ray collisions take place in interstellar helium.

   Compared to the number of p-nucleus cross sections measured,
the data on nucleus-nucleus cross sections are rather scanty.
Cumming et al. (1974) have measured several spallation cross
sections of Cu induced by nitrogen ions at a total kinetic energy
3.9 GeV (i.e., at 280 MeV/u). Here, and in the subsequent discussion we adopt the notation MeV/u to denote the energy per
atomic mass unit (or, approximately, energy per nucleon).

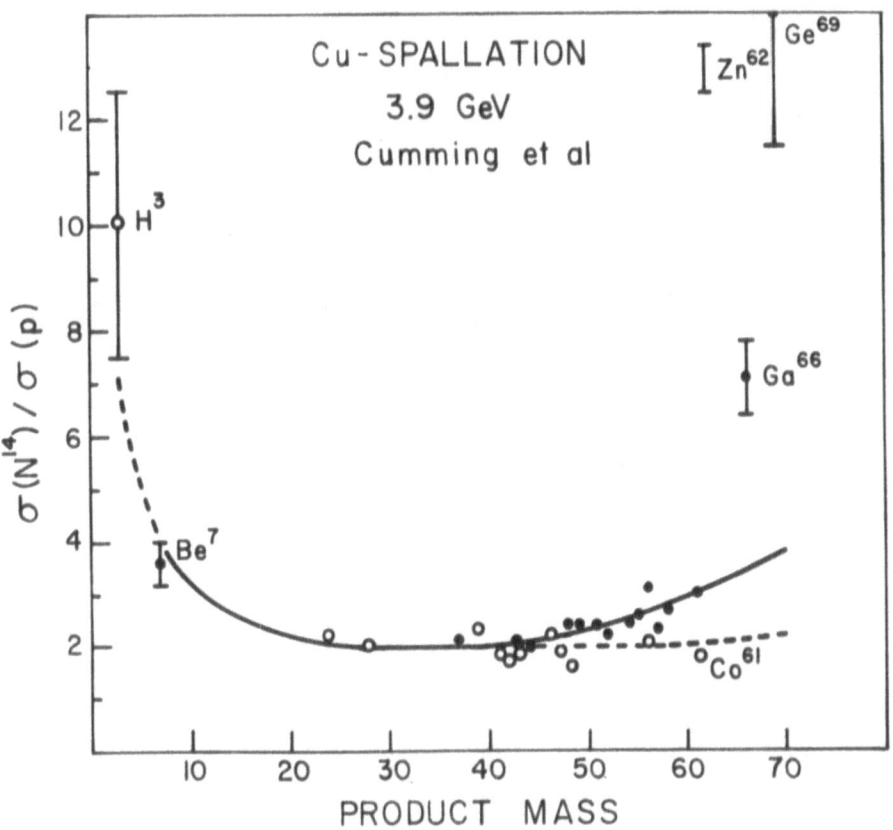

Figure 2. The relative values of partial cross sections of
N-Cu and P-Cu as a function of product mass. The data are
those of Cumming et al. (1974). The absolute normalization
of the two sets of data is uncertain.

Cumming et al. find that, to a first approximation, the partial cross sections of ^{14}N with Cu are proportional to those of p-Cu reactions at 3.9 GeV. The ratios of ^{14}N to p-induced partial cross sections measured by Cumming et al. (1974) are shown in Figure 2. Certain deviations from proportionality to p-Cu interactions can be seen: (1) light fragments like ^{3}H and ^{7}Be are produced at a higher rate, and (2) the yields of neutron deficient products with mass numbers A > 35 are enhanced (shown by the black circles).

Karol (1974) has studied the collisions of helium nuclei at 720 MeV kinetic energy per nucleus (180 MeV/u) with copper. He also finds the partial cross sections to be proportional to those of p-Cu reactions (with a scaling factor of 1.9). However, for ^{4}He-induced reactions, neutron-deficient products with A > 35 are suppressed, unlike those of nitrogen induced reactions.

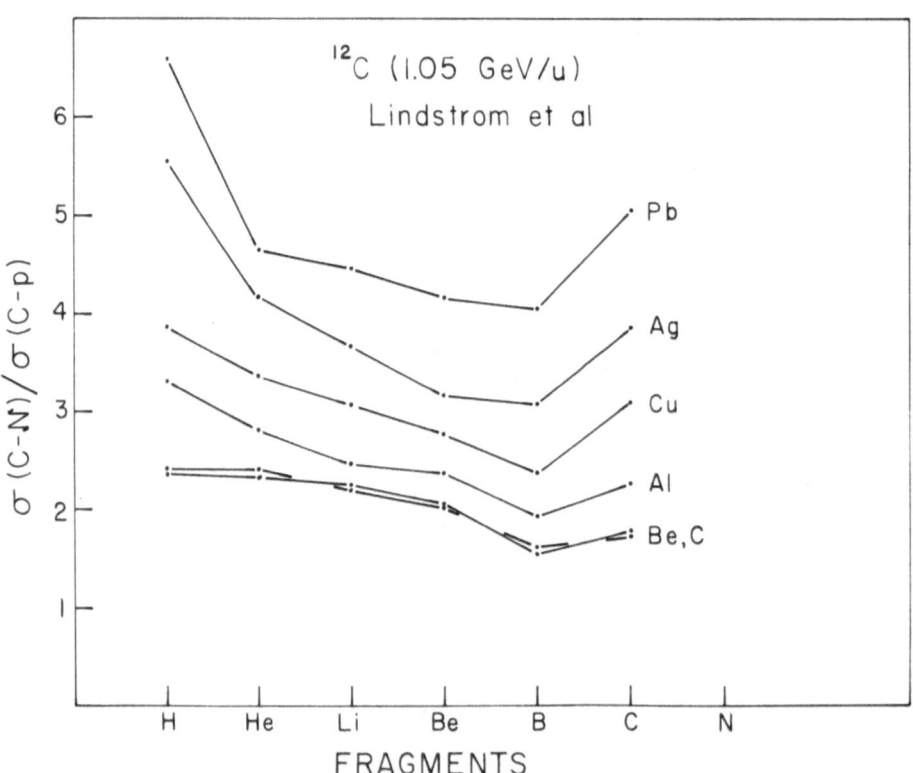

Figure 3. Ratios of C-nucleus to C-p cross sections at 1 GeV/u, as a function of product elements. The data are based on Lindstrom et al. (1975).

Lindstrom et al. (1975) have recently measured the partial cross sections for the breakup of C and O nuclei in collisions with H, Be, C, Al, Cu, Ag, and Pb. Figure 3 shows a comparison of C-nucleus interactions with those of C-p reactions. Again we note a general proportionality between C-nucleus and C-p reactions. The deviations from complete proportionality have been explained by Lindstrom et al. in terms of (1) nuclear transparency (the energy deposition is less in p-nucleus interactions; so, as ΔA increases, the relative yields of p-nucleus reactions diminish progressively), and (2) giant dipole resonance, as a result of which single-nucleon stripping, e.g., of ^{12}C is enhanced in collisions with heavy nuclei.

As a starting point for calculating nucleus-nucleus cross sections, we take the p-nucleus cross sections, and determine the appropriate scaling factors and other correction factors by which the latter should be multiplied.

We denote by N_1 the nuclide whose breakup mode is being explored; it may be a stationary "target," or it may be a beam of particles. In the former case one determines the yields of radioactive products in the target, while in the latter, one identifies the products from the fragmented beam particles. The "collision partner" with which N_1 collides is denoted by N_2.

The cross sections of N_1-N_2 reactions can be calculated from those of N_1-p interactions, σ (N_1-p) with the equation:

$$\sigma\ (N_1\text{-}N_2) = \sigma\ (N_1\text{-}p)\ S_c\ \epsilon_n\ \epsilon_L\ \epsilon_1\ \epsilon_\Delta \tag{3}$$

Here S_c represents the scaling factor; it is a function of nuclear skin thickness, as discussed by Lindstrom et al. (1975). We give here a simple prescription for S_c, based on empirical fitting. The factors ϵ_n, ϵ_L, ϵ_1, and ϵ_Δ represent correction factors for neutron-deficient products, for light products, for single-nucleon stripping and for reactions with large ΔA, respectively. Tables 2 to 4 give the expressions for these factors.

Table 2 represents the appropriate parameters for collisions with ^4He. The definitions of the expressions S, T, Z and M in Tables 2 and 3 are given in papers ST. For collisions with ^4He, equation 3 may be used down to a kinetic energy of 800 MeV (200 MeV/u), using the value of σ (p-N_1) at the same total kinetic energy, (not at same energy per nucleon). The restriction to $Z_1 > 8$ in the expression ϵ_L in Table 2 is based on the recent work of Raisbeck and Yiou (1975).

Tables 3 and 4 are for collisions with relatively light collision partners (Li to F) and heavier ones (Ne to Pb),

TABLE 2

**The Parameters of Eq. 3 for the Breakup of
Nuclei N_1 (Be to Zn) on ^{4}He.**

S_c = 1.8; for $\Delta A \leq 2$ and $Z_1 \leq 8$, set S_c = 1.6

$$\epsilon_n = \begin{cases} \left[1-0.15\ (A_z-A-1)\right] \geq 0.3 \text{ for } 35 < A < A_z-1 \\ 1 \qquad\qquad\qquad\qquad \text{otherwise} \end{cases}$$

$A_z = \left[S-(S^2-4\ TZ)^{\frac{1}{2}}\right]/2T\ -M+1$

$\epsilon_L = 1 + 0.4\left[1 + 0.02(Z_1/Z)^2\right](1-1.5\ Z/Z_1)$ for $3 \leq Z \leq 5$.

If $Z_1 > 8$, set Z_1 equal to 8.

$\epsilon_1 = 1$

$\epsilon_\Delta = 1$

---

respectively. Equation 3 is considered to be applicable above
kinetic energies per nucleon of 600 MeV/u, i.e., total kinetic
energies > 3.6 GeV; the cross sections are calculated with the
asymptotic high-energy values ($E > E_0$ of papers ST). The introduction of the expression ϵ_Δ is due to Meyer and Cassé (1975).
However, the exact value of the parameter Q of ϵ_Δ is uncertain;
the energy independence of the fragmentation parameters
$P_{VH,M}$ and $P_{VH,L}$, observed by Cleghorn et al. (1968) suggests
$Q \approx 0.0$, while the numerical values of the fragmentation parameters of the VH nuclei (atomic numbers 20 to 28) with carbon or
with air-like nuclei in emulsions suggest $Q \approx 0.3$. We tentatively
adopt Q = 0.2 ± 0.1. However, experiments on the spallation of
Fe or Cu in collisions with nuclei like C, N or O are required
at energies $\gtrsim$ 1 GeV/u to establish this value with adequate
precision. A better determination of Q is essential for calculating the propagation and interactions in the atmosphere of
cosmic-ray nuclei like iron.

TABLE 3

The Parameters of Eq. 3 for the

Breakup of Nuclei N_1(Be to Zn) on Collision Partners N_2(Li to F)

$S_c = 2(1 - 0.22 \ QA_1)^*$

$\epsilon_n = \begin{cases} 1 + 0.011 \ (A - 35) \ \text{for} \ 35 < A < A_z^\dagger \ \text{and} \ A_2 \geq 6 \\ 1 \qquad\qquad\qquad \text{otherwise} \end{cases}$

$\epsilon_L = \quad 1 + 0.4 \ [1 + 0.02 \ (Z_1/Z)^2](1 - 1.5 \ Z/Z_1), \ \text{for} \ 3 \leq Z \leq 5, \ \text{and} \ Z_1 \leq 10$

$\epsilon_L = \quad \text{larger of} \ \begin{cases} 1 + 0.4[1 + 0.02(Z_1/Z)^2](1 - 1.5 \ Z/Z_1) \ (1 - 0.22 \ Q \ A_1)^{-1} \\ \epsilon_\Delta \end{cases}$

$\qquad\qquad \text{for} \ 3 \leq Z \leq 5, \ \text{and} \ Z_1 > 10$

$\epsilon_1 = 1$

$\epsilon_\Delta = \exp[Q(A_1 - A)]$

$Q \ = 0.02 \pm 0.01$

---

*for $A_1 > 35$, $S_c = 2(1 - 0.22 \ Q \ A_1)[1 + \dfrac{0.011}{3}(A_1 - 35)]^{-1}$, to compensate

for enhancement of n-deficient products.

$^\dagger A_z$ is defined in Table 2.

There are no adequate experimental data for constructing
equations for the breakup of nuclei like Ne, Mg, Si and Fe when
they collide with heavy collision partners (heavier than fluorine).

2.4 Monte Carlo calculations

This topic is treated more extensively by Bertini (1975) in the
current volume. He cited instances in which the Monte Carlo
calculations are likely to yield more precise results than the
semiempirical calculations. There are additional cases when
Monte Carlo calculations should be used: (1) when it is essential
to know the distributions in angle and energy for the ejected
nucleons, (2) when the nuclear reaction is induced by neutrons,
and (3) when the particles have relatively low energies
($E \leqslant 60$ MeV).

TABLE 4

The Parameters of Eq. 3 for the Breakup of Nuclei N_1 (Be to Ne)
Against Collision Partners N_2 (Ne to Pb)

$$S_c = \left[1.6 + 0.07A_2^{2/3}\right](1-0.22\ QA_1)$$

$$\epsilon_L = \left\{1 + 0.4\left[1 + 0.02\left(\frac{Z_1}{Z}\right)^2\right]\left(1-1.5\ \frac{Z}{Z_1}\right)(2.71/A_2^{1/3})\right\}(1-0.22\ QA_1)^{-1}$$

for $3 \leq Z \leq 5$

$$\epsilon_1 = \begin{cases} 1 + 0.04\ (Z_2-10) & \text{for } \Delta A = 1 \\ 1 & \Delta A > 1 \end{cases}$$

$$\epsilon_n = 1$$

$$\epsilon_\Delta = \exp\left[Q(A_1 - A)\right]$$

$$Q = 0.02 \pm 0.01$$

The spallation cross sections of iron are of considerable interest; e.g., iron is quite abundant in cosmic rays, some particle detectors have support structures of iron, and iron is also an important component of meteorites.

We present in Table 5 the cross sections from the Monte Carlo calculations for p-Fe interactions at 150 MeV/u and 370 MeV/u. The programs were provided to us by Friedlander at Brookhaven (Chen et al. 1968a, 1968b, 1971, and Dostrovsky et al. 1959). At 150 MeV/u these are based on 1000 VPOT cascade calculations and about 5000 DFF evaporation calculations, and at 370 MeV/u on 1000 VEGAS STEPNO cascades and about 5000 DFF evaporation calculations. We used the particular version of VPOT that prohibits nucleons from approaching closer than 0.5 Fermi to each other. For comparison, available experimental values are given in parentheses.

TABLE 5

Spallation Cross Sections of p-Fe Interactions
from Monte Carlo Calculations[*]

| Product | | Cross Sections (in mb) at | |
Element	Mass Number	150 MeV/u	370 MeV/u
Co	56	11 ± 4	$0.4 ^{+0.7}_{-0.4}$ (0.96)[*]
	55	10 ± 4	1.4 ± 1 (0.74)
	54	$0.7 ^{+0.9}_{-0.7}$	$0.8 ^{+1}_{-0.8}$
	53	$0.4 ^{+0.7}_{-0.4}$	< 0.5
	52	< 0.5	$0.1 ^{+0.4}_{-0.1}$
Fe	55	95 ± 11 (110 ± 10)	66 ± 9 (64 ± 4)
	54	53 ± 8	39 ± 7
	53	31 ± 6 (30 ± 2)	22 ± 5
	52	2.1 ± 1.7	1.7 ± 1.5
	51	$1.1 ^{+1.2}_{-1.1}$	$0.8 ^{+1.0}_{-0.8}$
Mn	55	27 ± 6	31 ± 6
	54	40 ± 7 (38 ± 8)	36 ± 7 (34 ± 5)
	53	33 ± 7	31 ± 6
	52	57 ± 9 (14 ± 3)	38 ± 7 (18)
	51	17 ± 5 (5.8 ± 1.2)	13 ± 4 (4)
	50	3 ± 2	2.3 ± 1.8
	49	$0.4 ^{+0.7}_{-0.4}$	$0.1 ^{+0.4}_{-0.1}$
Cr	54	4 ± 2	5 ± 3
	53	9 ± 3	10 ± 4
	52	17 ± 5	17 ± 5
	51	44 ± 8 (40 ± 10)	30 ± 6 (25 ± 4)
	50	35 ± 7	22 ± 5
	49	23 ± 5 (6.1 ± 1.7)	17 ± 5 (4.8)
	48	2.2 ± 1.7 (0.5 ± 0.1)	$0.8 ^{+1.0}_{-0.8}$ (0.7)
	47	$0.4 ^{+0.7}_{-0.4}$	$0.3 ^{+0.6}_{-0.3}$
V	53	< 0.5	$0.5 ^{+0.8}_{-0.5}$
	52	1.7 ± 1.5	2.0 ± 1.6
	51	4 ± 2	3 ± 2
	50	20 ± 5	14 ± 4

[*]Available experimental values are given in parentheses ( ).

TABLE 5 (cont.)

Product		Cross Sections (in mb) at	
Element	Mass Number	150 MeV/u	370 MeV/u
V (cont.)	49	22 ± 5 (33 ± 5)	14 ± 4 (31)
	48	29 ± 6 (15 ± 2)	21 ± 5 (9)
	47	8 ± 3 (5.9 ± 1.9)	5 ± 2 (2.4)
	46	1.3 ± 1.3	$0.7^{+0.9}_{-0.7}$
	45	$0.1^{+0.4}_{-0.1}$	$0.3^{+0.6}_{-0.3}$
Ti	51	$0.3^{+0.6}_{-0.3}$	$0.1^{+0.4}_{-0.1}$
	50	$0.4^{+0.7}_{-0.4}$	$0.7^{+1.0}_{-0.7}$
	49	6 ± 3	6 ± 3
	48	11 ± 4	10 ± 4
	47	20 ± 5	16 ± 5
	46	17 ± 5	19 ± 5
	45	7 ± 3 (4.5 ± 1)	11 ± 4 (3.7)
	44	$0.5^{+0.8}_{-0.5}$	2.0 ± 1.6
	43	< 0.5	$0.3^{+0.6}_{-0.3}$
Sc	49	< 0.5	$0.3^{+0.6}_{-0.3}$
	48	$0.4^{+0.7}_{-0.5}$	$0.7^{+1.9}_{-0.7}$ (0.35)
	47	$0.9^{+1.1}_{-0.9}$ (0.7 ± 0.2)	2.8 ± 1.9 (1.0)
	46	3.5 ± 2.2 (3.0 ± 0.6)	6 ± 3 (3.5)
	45	6 ± 3	6 ± 3
	44	6 ± 3 (5.9 ± 0.4)	12 ± 4 (2.6)
	43	1.4 ± 1.4 (2.5 ± 0.2)	8 ± 3
	42	$0.3^{+0.8}_{-0.3}$	$0.7^{+0.9}_{-0.7}$
Ca	47	< 0.5 (0.007)	$0.1^{+0.4}_{-0.1}$ (0.007)
	46	$0.1^{+0.4}_{-0.1}$	$0.8^{+0.8}_{-0.8}$
	45	$0.1^{+0.4}_{-0.1}$ (0.36)	$1.2^{+1.3}_{-1.2}$ (0.56)
	44	$0.5^{+0.8}_{-0.5}$	2.8 ± 1.9
	43	1.6 ± 1.4	4.6 ± 2.5
	42	$1.1^{+1.2}_{-1.1}$	6 ± 3
	41	$0.7^{+0.9}_{-0.7}$	5.0 ± 2.6
	40	< 0.5	$0.5^{+0.8}_{-0.5}$
K	43	< 0.5 (0.11)	1.3 ± 1.3 (0.4)
	42	$0.4^{+0.7}_{-0.4}$ (0.25)	2.7 ± 1.9 (0.7)
	41	$0.1^{+0.4}_{-0.1}$	3.2 ± 2.0
	40	1.4 ± 1.4	6 ± 3
	39	$0.5^{+0.8}_{-0.5}$	5.4 ± 2.7
	38	< 0.5	$0.5^{+0.8}_{-0.5}$

TABLE 5 (cont.)

Product		Cross Sections (in mb) at	
Element	Mass Number	150 MeV/u	370 MeV/u
Ar	42	< 0.5	$0.5 \pm {}^{0.9}_{0.5}$
	41	< 0.5	$0.3 \pm {}^{0.6}_{0.3}$
	40	$0.1 \pm {}^{0.4}_{0.1}$	$1.2 \pm {}^{1.3}_{1.2}$
	39	< 0.5	1.9 ± 1.6 (4.1 ± 0.6)
	38	$0.1 \pm {}^{0.4}_{0.1}$ (~ 0.3)	3.5 ± 2.2 (~ 6)
	37	< 0.5 (0.19)	$0.4 \pm {}^{0.7}_{0.4}$ (3.3)
Cl	39	< 0.5 (0.024)	$0.1 \pm {}^{0.4}_{0.1}$ (0.045)
	38	< 0.5	$1.1 \pm {}^{1.5}_{1.1}$ (0.17)
	37	< 0.5	$0.1 \pm {}^{0.4}_{0.1}$
	36	< 0.5	1.6 ± 1.5
	35	< 0.5	$0.9 \pm {}^{0.9}_{0.9}$
	34	< 0.5 (0.11 ± 0.04)	$0.1 \pm {}^{0.4}_{0.1}$ (0.11)
S	37	< 0.5	$0.3 \pm {}^{0.6}_{0.3}$
	36	< 0.5	$0.5 \pm {}^{0.8}_{0.5}$
	35	< 0.5 (0.18 ± 0.09)	1.3 ± 1.3 (0.23)
	34	< 0.5	< 0.5
	33	< 0.5	$0.3 \pm {}^{0.6}_{0.3}$
	32	< 0.5	$0.1 \pm {}^{0.4}_{0.1}$
P	35	< 0.5	$0.1 \pm {}^{0.4}_{0.1}$
	34	< 0.5	$0.1 \pm {}^{0.4}_{0.1}$
	33	< 0.5 (0.065)	< 0.5
	32	< 0.5 (0.2 ± 0.1)	$0.8 \pm {}^{0.8}_{0.8}$ (0.5 ± 0.2)
	31	< 0.5	$0.9 \pm {}^{0.9}_{0.9}$
Si	32	< 0.5	$0.1 \pm {}^{0.4}_{0.1}$
	31	< 0.5 (0.026 ± 0.013)	$0.1 \pm {}^{0.4}_{0.1}$ (0.12)
	30	< 0.5	$0.1 \pm {}^{0.4}_{0.1}$
	29	< 0.5	< 0.5
Al	28	< 0.5	< 0.5
	27	< 0.5	< 0.5
	26	< 0.5	$0.1 \pm {}^{0.4}_{0.1}$

3.   APPLICATIONS TO COSMIC RAY PHYSICS

     This topic is treated in the present volume by Raisbeck and
Yiou (1975). We shall complement their presentation by emphasiz-
ing the latest experimental and theoretical studies presented at
the 14th International Cosmic Ray Conference, August 1975, in
Munich.

     We shall first briefly review the evidence that cosmic rays
have undergone extensive spallation reactions since their acceleration in source regions. In Figure 4, the composition of
cosmic rays in the vicinity of the earth, at energies between 1

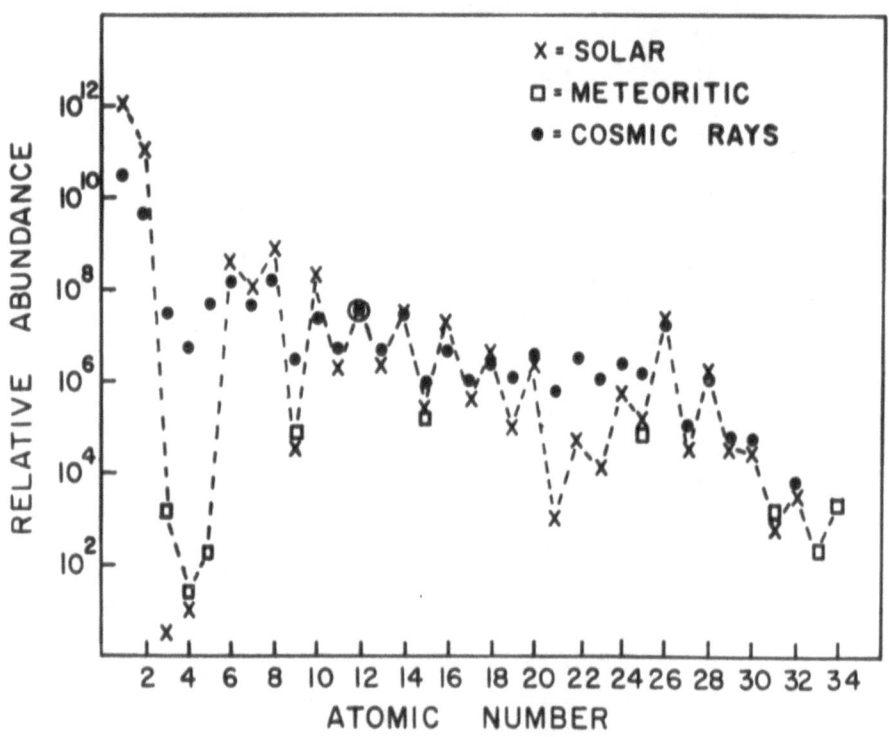

Figure 4. A comparison of the abundances of elements in cosmic
rays and in the solar system. The solar hydrogen abundance is
normalized to 10^{12}, and the cosmic-ray abundances are normalized
to the solar values at magnesium. The solar values of Li and Be
are lower than the meteoritic, because these elements have burned
up in the sun. For solar Ne and Ar, the data of Geiss (1975) on
the composition of the solar wind was used.

and 10 GeV/u, is compared with that of the sun. (The solar composition is similar to that of normal stars, and for the non-volatile component, also similar to that of the carbonaceous chondrite meteorites.) The light (L) elements Li, Be and B are extremely scarce in nature; this is due to their easy consumption in nucleosynthetic processes. In cosmic rays these elements are rather abundant, due to their creation by spallation of heavier elements. A somewhat similar disparity occurs for the elements between sulphur and iron. The abundance of these spallation products is consistent with that of the light (L) elements produced in traversing a path length of 5.5 g/cm^2 of interstellar material (assumed to contain 10% helium by number). The abundances of other nuclear interaction products (^{2}H, ^{3}He, e$^+$) in cosmic rays support this interpretation.

The cosmic-ray abundances of the elements from lithium to nickel that arrive near the earth have now been measured for all individual elements. Table 6 shows these abundances above the atmosphere at rigidities R ≥ 4 GV. This table is based on

TABLE 6

Cosmic-ray Abundances Above the Atmosphere

at Rigidities R ⩾ 4 GV

Li	18 ± 2	Mg	19 ± 1	Sc	0.4 ± 0.2
Be	10.5 ± 1	Al	2.8 ± 1	Ti	1.7 ± 0.3
B	28 ± 1	Si	14 ± 2	V	0.7 ± 0.3
C	100*	P	0.6 ± 0.2	Cr	1.5 ± 0.4
N	25 ± 2	S	3 ± 0.4	Mn	0.9 ± 0.2
O	91 ± 2	Cl	0.5 ± 0.2	Fe	10.8 ± 1.4
F	1.7 ± 0.4	Ar	1.5 ± 0.3	Co	0.05 ± 0.02
Ne	16 ± 2	K	0.8 ± 0.2	Ni	0.5 ± 0.1
Na	2.7 ± 0.4	Ca	2.2 ± 0.5		

*Normalization: 100 for carbon

Table 3 of Shapiro and Silberberg (1975a), who provide pertinent
references to the experimental data. However, we give new values
for Co on the basis of recent measurements by Juliusson and Meyer
(1975), and Ormes et al. (1975), and for Ni (on the basis of
Benegas et al. 1975a, Ormes et al. 1975, and Lund et al. 1975).

3.1 Isotopic composition, effects of spallation

Due to the breakup of many nuclei, the isotopic composition
of cosmic rays arriving at the top of the atmosphere differs from
that of the general isotopic abundances of elements in nature
(Tsao et al., 1973).

Figure 5 shows the isotopic abundances of the cosmic rays,
calculated from the elemental composition at the sources by
Shapiro et al. (1975). The isotopic ratios for a given element
at the source were taken to be the same as those estimated for
ordinary matter by Cameron (1973). In Figure 5 the cross-hatched
columns give the abundances at the sources. The other columns
represent the arriving composition: the white areas denote the
surviving source components and the black areas the secondary
(spallation) products. While the "alpha-particle" nuclides ^{12}C,
^{16}O, ^{20}Ne, ^{24}Mg, ^{28}Si, ^{32}S, ^{40}Ca and ^{56}Fe reflect the processes
of nucleosynthesis, most of the other abundances reflect the
spallation process. In Figure 5 we have omitted (for reasons of
space) most of the nuclides with mass numbers between 34 and 56
which are predominantly formed by spallation; only those with
significant source components are plotted.

The experimental study of isotopes heavier than helium is
still in its infancy. Some pioneering work in this field was
done by Jönsson et al. (1970) who were able to resolve the
isotopes of carbon, and Webber et al. (1973a) who studied the
isotopes of the set of elements Li to N.

We have calculated the isotopic composition of elements Li
to Fe (Tsao et al., 1973). For the lighter elements, a comparison is now possible with measured values, as shown in Table 7.
The measured values for Li, Be and B are based on Garcia-Munoz
et al. (1975a), and for Be and B also on Hagen et al., (1975).
Thus far, results on ^{10}Be are discordant (see footnote to
Table 7). The values for C come from Bjarle et al. (1975) and
Fisher et al. (1975), for N from Dwyer and Meyer (1975), and
for O from Maehl et al. (1975), supported by Jacobsson et al.
(1975).

Isotopic measurements on cosmic rays with $Z > 8$ are in
conflict: Fisher et al. (1975) find an appreciable amount of
^{22}Ne, while Juliusson (1975) does not. Webber et al. (1973b)
finds that $\sim 1/3$ of iron is ^{58}Fe, and 1/3 is ^{54}Fe, while

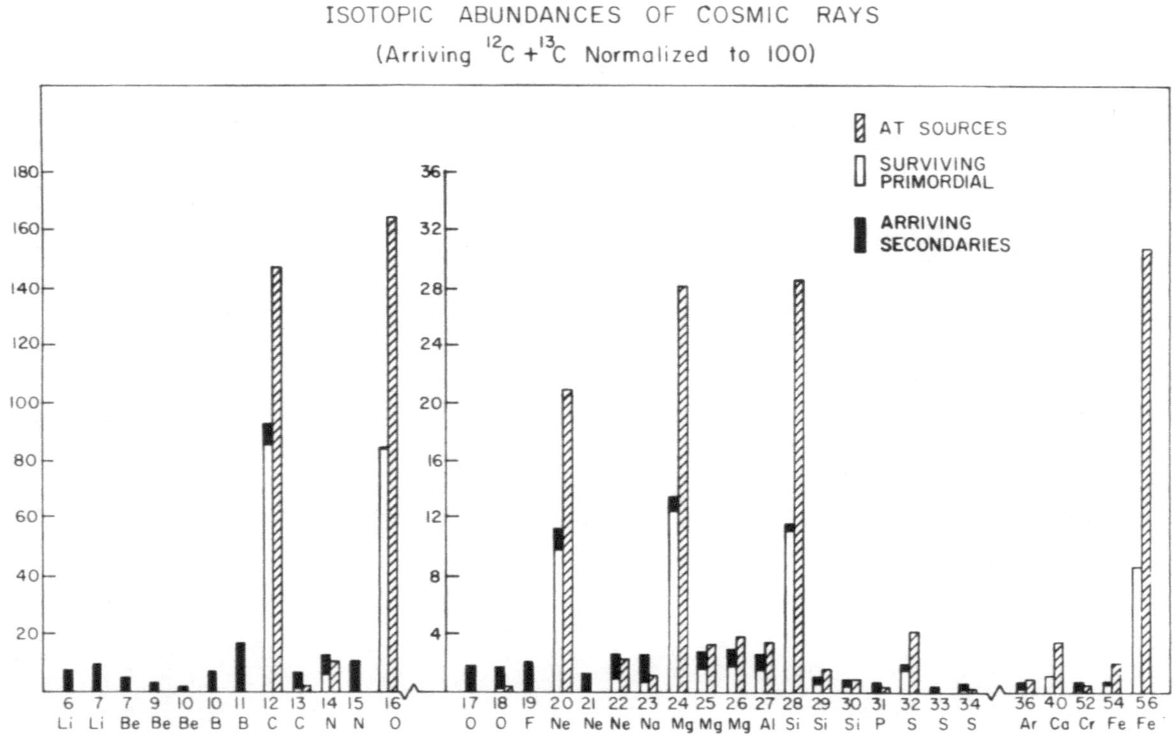

Figure 5. The relative abundances of nuclides at cosmic ray sources and near the earth. All abundances have been normalized to a value of 100 for the arriving two isotopes of carbon. In the "arriving" columns, the secondaries and the surviving source component are shown separately. The arriving abundances are shown at comparable rigidities 4 ≳ GV.

TABLE 7

Comparison of Measured and Calculated

Isotopic Abundances of Arriving Cosmic Rays

Isotope	Measured Abundance (%)	Calculated[†] Abundance (%)
^{6}Li	52 ± 8	56 ± 6
^{7}Li	48 ± 8	44 ± 6
^{7}Be	61 ± 6	60 ± 6
^{9}Be	30 ± 4	32 ± 3
^{10}Be	4 to 11[*]	8 ± 2[*†]
^{10}B	34 ± 3	32 ± 4[*]
^{11}B	66 ± 3	68 ± 4
^{12}C	93 ± 3	94 ± 1
^{13}C	7 ± 3	6 ± 1
^{14}N	55 ± 6	55 ± 5
^{15}N	45 ± 6	45 ± 5
^{16}O	> 91	96 ± 2
^{17}O	< 6.5	2 ± 1
^{18}O	2.5 ±1.5	2 ± 1

[*]Results on ^{10}Be are still controversial. **Hagen et al. (1975)** obtained 11 ± 5 %, while **Garcia-Munoz et al. (1975)** report 4^{+8}_{-4} % from IMP-7 observations, **and < 8 % from IMP-8** (the quoted errors correspond to 90 % confidence limits). The calculated value corresponds to decay of one-half of the ^{10}Be into ^{10}B.

[†]While the measured values are at energies 50 to 450 MeV/u, the calculated values are based on high-energy cross sections (E < 2000 MeV/u), because these are now the ones best known. Actually, the measured particles had higher energies, near 500 MeV/u at the time of collision, owing to adiabatic deceleration in the solar system and ionization losses in interstellar space. With the cross sections at ~ 500 MeV/u, the calculated abundance of ^{10}Be, for 50 % survival, is reduced to 7 ± 2 %.

Fisher et al. find that the abundance of ^{58}Fe is less than 10%. and ^{54}Fe about 20 ± 10%. Also Benegas et al. (1975b), Bartholomä et al. (1975) and Clapham et al. (1975) report appreciable amounts of ^{54}Fe. Whether the large ^{54}Fe abundance is real, or reflects poor resolution is still uncertain. Some skepticism is not unwarranted when we recall that a few years ago high values of the ratios Cr/Fe and (Li, Be, B)/(C, N, O) were reported in several successive experiments. An experiment that clearly resolves the isotopes of iron is required.

3.2 Information content of six classes of cosmic ray nuclides

The subsequent discussion can be followed more easily if we subdivide the cosmic-ray nuclides into six classes, each of which has a specific information content. These are displayed in Table 8.

3.2.1. The stable nuclei that originate in cosmic-ray sources can serve either as probes of (1) nucleosynthesis in these sources, or (2) special conditions for injection or acceleration of particles. Table 9 taken from Silberberg et al. (1975) displays the processes of nucleosynthesis that contribute to the buildup of the nuclides shown. For the nuclides shown in parentheses (e.g., ^{13}C) the secondary component is likely to be so large as to obscure the magnitude of the source component.

A comparison of the elemental abundances of cosmic rays (extrapolated back to the sources) with the general elemental abundances in the sun and normal stars is informative (Shapiro et al., 1975). Table 10 shows the relative abundance of elements at the cosmic ray sources, computed from the diffusion equation and appropriate cross sections. These values are used in Table 11 which presents the ratios of the elemental abundances of cosmic rays at the sources to the solar (or general) abundances. The two sets of abundances are normalized to unity for the elements Mg, Si and Fe; the ratio is also nearly unity for Na, Al, Ca and Ni, as well as for the ensemble of elements beyond nickel.

One interpretation of the departures from unity in Table 11 is based on nucleosynthesis: the composition of the cosmic rays is considered to reflect that of supernovae, i.e., hydrogen, and also the products of hydrogen-burning like He and N, are depleted in cosmic rays. A recent version of this interpretation is presented by Schramm et al. (1975). According to this model, some mixing of the supernova ejecta with the surrounding stellar envelope and interstellar medium takes place before the acceleration of cosmic rays some 100 years after the explosive nucleosynthesis in the supernova outburst.

TABLE 8

Six Classes of Cosmic-Ray Nuclides

Source components	Spallation products
Stable	Stable
Long-lived radioactive	Long-lived radioactive
Electron-capture nuclides*	Electron-capture nuclides*

*With positron-decay forbidden, or nearly forbidden

TABLE 9

Arriving Cosmic Rays as Probes of Nucleosynthesis in Cosmic Ray Sources

Nuclide	Process
^{12}C, ^{16}O	He-burning
(^{13}C), ^{14}N	CNO-cycle
^{22}Ne	N-burning in He-zone
^{20}Ne, (^{23}Na), $^{24-5-6}Mg$	Explosive C-burning
^{27}Al, $^{29,30}Si$, (^{31}P)	
^{28}Si, ^{32}S, ^{36}Ar, ^{40}Ca	Explosive O-burning
^{44}Ti?, (^{52}Cr), (^{55}Mn), $^{54,56-7}Fe$	Explosive Si-burning
^{59}Co, $^{58,60-1-2}Ni$ [^{59}Ni?]	
^{58}Fe, ^{64}Ni	n-rich, high density explosion
$Z \geq 30$	r, s-processes

TABLE 10

Calculated Source Abundances of Cosmic Rays with $Z \leq 28$

H[*]	2×10^4	Mg	24 ± 2	K	$\sim 0.1 \, ^{+\,0.5}_{-\,0.1}$
He	2600	Al	2.3 ± 1.5	Ca	2.5 ± 1
C[†]	100	Si	21 ± 3	Ti	$\sim 0.1 \, ^{+\,0.5}_{-\,0.1}$
N	7.4 ± 3[‡]	P	$0.2 \, ^{+\,0.4}_{-\,0.2}$	Cr	$0.5 \, ^{+\,0.9}_{-\,0.5}$
O	111 ± 2	S	3 ± 0.7	Mn	$\sim 0.1 \, ^{+\,0.5}_{-\,0.1}$
Ne	16 ± 3	Cl	$\sim 0.1 \, ^{+\,0.5}_{-\,0.1}$	Fe	22 ± 3
Na	0.9 ± 0.7	Ar	0.8 ± 0.6	Ni	1.0 ± 0.3

[*]Would be 5×10^4, for a spectrum that depends on velocity.

[†]Normalization.

[‡]The errors include uncertainties in cross sections, in the arriving abundances, in the mean path length, and in the distribution of path lengths. The latter two were estimated, respectively, from the 8 % error in the L/M ratio, and a comparison of the source abundances evaluated with a gaussian, a pure exponential, and our adopted exponential-like distribution of path lengths. Li, Be and B were assumed to be absent from the sources.

TABLE 11

Ratio of Cosmic-ray Abundance at the Sources to Solar Abundance[*]

Element	Ratio
H[†]	0.03
He	0.05
N	0.1
O	0.25
C	0.4
Ne, S, Ar	0.25 to 1
Na, Al, Ca, Ni	~ 1

[*]The ratio is normalized to unity for the set
of elements Mg, Si, and Fe.

[†]Based on a rigidity-dependent spectrum at the
sources (for a velocity-dependent spectrum,
the ratio would be ~ 0.1).

Some support for models of the above type is derived from
the composition of the trans-Ni elements: the elements formed
by rapid neutron capture (r-process), e.g., Pt, seem to be enhanced in cosmic rays relative to the s-process elements. There
are some (still unconfirmed) observations that suggest special
conditions of nucleosynthesis in cosmic rays sources: (a) preliminary data by Ormes et al. (1975) suggest an overabundance of
^{22}Ne, which is formed by N-burning in a He-zone. But Juliusson
(1975) does not find it. (b) Measurements of Cartwright et al.
(1971) (unlike those of Webber et al., 1972, at higher energies)
imply that ^{23}Na and ^{27}Al are abundant at the sources; this could
come about in an explosive C-burning environment somewhat enriched
in neutrons. (c) Webber et al. (1973b) find that iron nuclides
have a large spread of mass values, from 54 to 58. [Fisher et al.
(1975) find appreciably less, if any ^{58}Fe.) Modifications in

silicon burning, e.g., a more n-rich environment, would be in-
voked to interpret Webber's results.

Alternate explanations of the cosmic-ray abundances shown
in Table 11 start out with thermal source abundances like those
in the general composition of the elements. However, this initial composition is considered to be affected by the mechanism
of injection into the acceleration region. Havnes (1971) has
proposed a dependence of cosmic-ray composition on the first ionization potential. Elements with a high ionization potential
(whose atoms tend to remain neutral) are discriminated against in
the acceleration region. This theory has been further elaborated
by Cassé et al. (1975b) who discuss physical conditions in which
such a process could occur. Kristiansson (1971) has suggested a
dependence on ionization cross sections. Acceleration processes
that depend on the charge of the nucleus have been considered by
Cowsik and Wilson (1973).

The enrichment of r-process elements and the low value of
nitrogen in cosmic rays (at the sources) favors the interpretation
based on nucleosynthesis, but not definitively. Conclusive tests
are discussed in subsections 3.2.2 and 3.2.3: models of nucleosynthesis imply recent generation of the nuclei, ($\leqslant 10^7$ years ago)
and acceleration soon (in $\leqslant 10^3$ years) after nucleosynthesis.
The models dependent on injection imply a time scale $\geqslant 10^9$ years.

3.2.2. The long-lived radioactive nuclides that originate in
cosmic ray sources can be used to determine the time T_t that has
passed since the nucleosynthesis of cosmic-ray nuclides. A value
of $T_t \leqslant 10^7$ years implies an origin in recent supernovae or in
their shells, with cosmic-ray confinement principally in the
galactic disc. A value of $10^7 \leqslant T_t \leqslant 10^8$ would imply a similar
origin, but confinement in the halo. If $T_t \gg 10^8$ years, an
origin in old interstellar gas, or in extragalactic sources is
implied. The abundances of the transbismuth nuclides are particularly suitable for the determination of T_t.

The large abundance of nuclides with $Z \geq 90$ relative to the
Pb group or the Pt group has been interpreted as implying the
survival of some transuranic nuclides (including the four decay
branches), and thus a recent nucleosynthesis $T_t \leq 10^7$ years
(Israel et al. 1975). However, since the presence of ^{237}Np and
^{235}U has not been directly determined, the inference drawn is to
be regarded as suggestive rather than definitive.

Blake and Schramm (1973) have shown that the U/Th ratio
provides a sensitive test; it has a value of 5 to 10 for $T_t = 10^6$
or 10^7 years and 1 for $T_t = 10^9$. But experimental charge resolution is not yet adequate for the measurement of this ratio.

Of the transbismuth cosmic-ray particles, $\sim 1/3$ might be transuranic, but the fraction could be appreciably less if the possible effects of ionization losses on identification of particles at lower energies ($E \lesssim 1$ GeV/u) are taken into account. According to Blake and Schramm (1974) the ratio trans-U/(Th + U) ≈ 1 for $10^5 \leq T_t \leq 10^6$ years, and decreases for larger values of T_t. The data are consistent with $T_t \approx 10^7$ or more if the energy of particles—and consequently their atomic numbers—should be overestimated.

A recent summary of the data (including those from Skylab) has been given by Price and Shirk, (1975). Experimental progress during the next five years should permit the establishment of a definite value for T_t, and resolve the astrophysical problems discussed at the beginning of this subsection.

3.2.3. <u>The nuclides that decay only by electron capture and originate in cosmic-ray sources</u> permit a determination of the time T_a between nucleosynthesis and acceleration of cosmic-ray nuclei (Cassé and Soutoul, 1974). In the synthesis of elements certain nuclides are produced that normally would decay by electron capture. If, however, they are accelerated prior to decay, they will survive in cosmic rays. These nuclides can thus be used to determine whether the primordial cosmic-ray "material" originates in supernovae or in old interstellar gas. The estimation of the time T_a (employing the ratio ^{44}Ti/^{46}Ti) also permits a discrimination between different models of acceleration in supernovae (Kulsrud et al. 1972, with $T_a \approx 1$ to 3 years or Schramm et al. 1975, with $T_a \approx 100$ years). Methods for determining the time T_a are outlined in Table 12. These were proposed by Silberberg (1974). The <u>elemental</u> ratio Co/Ni can reveal whether $T_a > 1$ year. To learn whether $T_a > 50$ years, it would be useful to measure ^{44}Ti/^{46}Ti.

Recently Juliusson and Meyer (1975) measured the ratio Co/Ni and found it to be ~ 0.15 and two standard deviations below 0.4; Ormes et al. 1975, found it to be ≤ 0.14. Hence, Shapiro and Silberberg (1975b) and Soutoul et al. (1975) have concluded that ^{57}Co has decayed, and that cosmic rays must have been accelerated at least one year after nucleosynthesis.

3.2.4. <u>The stable secondary nuclei that are formed by spallation during propagation have been used</u> to determine the mean path length of material traversed. Many items of interest under this topic are treated by Raisbeck and Yiou in Sections 2 and 3.1. of their article in the current volume. (Some of the matter may be in source regions; we shall return to this topic in subsection 3.2.6.) We noted in Figure 5 and in Section 3.1. that among the

TABLE 12

The Ratio Co/Ni and the Abundance of ^{44}Ti versus the
Time T_a between Nucleosynthesis and Acceleration

T_a (years)	Conditions determining Co and ^{44}Ti abundances	Expected* Ratio Co/Ni	Expected* Ratio ^{44}Ti/^{46}Ti
$\ll 1$	Original ^{57}Co survives	0.4	0.10
$1 \ll T_a \ll 50$	^{57}Co decays, secondary Co dominates, ^{44}Ti survives	0.06	0.10
$50 \ll T_a \ll 10^5$	Co and ^{44}Ti predominantly secondary	0.06	0.04
$\gg 10^5$	^{59}Co from decay of ^{59}Ni, some secondary Co	0.10	0.04

*The "expected" abundances of ^{57}Co and ^{44}Ti at the sources are
based on Cameron's (1973) values of ^{57}Fe and ^{44}Ca.

nuclides with mass numbers 6 to 56, only the so-called "alpha-
particle" nuclides have minute fractions of secondary components.
Among the trans-Ni nuclei, the elements just lighter than those
at the r- and s-process abundance peaks ($43 \leq Z \leq 49$ and
$61 \leq Z \leq 75$) are likely to have large secondary components, and
also those with $79 \leq Z \leq 81$ lying between the Pt and Pb peaks.

The abundance of the stable secondary nuclides has not only
been used to determine the path length of material traversed by
cosmic rays, but also the distribution function of path lengths.
The mean path length is about 5.5 g/cm^2 for cosmic rays at
energies of 1 to 10 GeV/u, assuming that 10% of the atoms in the
medium are helium. The experimental data on the L/M [i.e.,
(Li, Be, B)/(C, N, O)] and $17 \leq Z \leq 25$/Fe ratios require an exponential-like distribution, i.e., many short path lengths and
some particles with large path lengths. The energy dependence
of the L/M and other secondary/primary abundance ratios also
require an energy dependence of the path length traversed, with
smaller mean path lengths for high-energy particles (at several
tens or 100 GeV/u). [Due to differences in cross sections and
energy dependence of path length, the Fe/(C + O) ratio also varies
with energy; recent data on this subject are presented by Orth
et al. (1975) and Matsubayashi et al. (1975).] Various modifications of this basic distribution have been proposed, among

them that of Golden et al. (1975). Cowsik and Wilson (1975) have considered the "nested leaky box" model, with some confinement of low-energy particles in cosmic-ray source regions, and a mean path length of 1.9 g/cm^2 at 100 GeV/u. A more drastic modification of the exponential-like distribution has been proposed by Rasmussen and Peters (1975); they consider the leakage rate of cosmic rays from the galaxy to be negligibly slow; only the nuclear breakup reactions contribute to making the distribution exponential-like. To compensate for the introduction of more particles with long path lengths, they introduce additional particles with short path lengths, from a nearby cosmic-ray source.

A good recent presentation of propagation calculations employing Laplace transforms has been given by Meneguzzi et al. (1975).

In this paper we shall not dwell upon the ^{2}H and ^{3}He nuclei, which are also secondary. Their abundances are measured with respect to ^{4}He. The effects of the low-energy anomalous ^{4}He component which may be produced locally, at the outskirts of the solar system (Fisk et al., 1974) makes the interpretation of the data below 100 MeV/u less certain and more difficult. Additional difficulties are introduced by the adiabatic deceleration as a function of the phase of the solar cycle (Meyer, 1975). Potentially, the data on ^{2}H and ^{3}He could be valuable in helping to determine the form and slope of the cosmic ray energy (or rigidity) spectrum prior to solar modulation; some recent estimates are given by Mewaldt et al. (1975).

3.2.5. <u>The long-lived radioactive secondary nuclei that are formed by spallation during propagation</u> permit an estimation of the confinement time T of cosmic rays in the galaxy. Section 3.2. of Raisbeck and Yiou treats the above topic in this volume. They discuss the cosmic-ray clocks ^{10}Be, ^{26}Al and ^{36}Cl, and the effects of a set of non-homogeneous media on the estimate of confinement time.

Current studies have concentrated on the nuclide ^{10}Be, but there are still experimental problems with the identification of beryllium isotopes: we pointed out in the footnote of Table 7 the disparate estimates of ^{10}Be/Be by Garcia-Munoz et al. (1975) and Hagen et al. (1975), which differ by a factor of 3. On the basis of certain considerations outlined below, the discrepancy can be reduced to a factor of ~ 2. Then the two experiments are not mutually inconsistent if allowance is made for statistical fluctuations. The mean value of energy/nucleon is higher by ~ 200 MeV/u for Hagen et al. (1975). Hence, (a) the larger relativistic factor $\beta\gamma$ of Hagen et al. permits a slightly higher value of ^{10}Be/Be; (b) the energy dependence of the path length, hence of the confinement time, works in the same direction;

(c) at low velocities, solar modulation, according to Parker (1963), has the form $\exp(-K/\beta)$ or $\exp(-K/R^n\beta)$ with $n < 1$, while at somewhat higher energies, it is more rigidity dependent: $[\exp(-K/R\beta)]$; the latter form of modulation admits a larger fraction of ^{10}Be into the solar system; (d) the higher yield of the fraction ^{10}Be/Be in nuclear reactions at higher energies contributes another increment in the same direction.

The estimated confinement times (assuming diffusion in a homogeneous medium) are $(3 \, {}^{+5}_{-2}) \times 10^6$ years, according to Hagen et al. (a similar value was obtained by Webber et al. 1973a), and $\sim 20 \times 10^6$ years, according to one estimate by Garcia-Munoz et al. (1975b).

Because isotopic resolution of ^{10}Be has not yet yielded convergent values, O'Dell et al. (1975) have explored again the Be/B ratio, using newly measured cross sections like those of Lindstrom et al. (1975). The latter report errors in partical cross sections of about 10%, with systematic components of about 6%. O'Dell et al. made use of experimental data with a rigidity cutoff, which provides a "filter" that enhances the ^{10}Be component relative to ^{7}Be. The calculated isotopic yields were accordingly multiplied by $(A_i/2Z_i)^u$, where i denotes the isotope and $u = 1.5$ the exponent of the integral cosmic-ray rigidity spectrum between 4 and 10 GV. The measured Be/B ratio (0.39 ± 0.025) is consistent with the calculated ratio for ^{10}Be survival (0.401 ± 0.050) and exceeds considerably the value for ^{10}Be decay (0.278 ± 0.034). If a homogeneous medium is assumed, the comparison of the latter with the experimental value implies a confinement time $T < 10^7$ years within one standard deviation and $T < 5 \times 10^7$ years within 2 standard deviations.

These upper limits are, in our view, not thus far clearly contradicted by the isotopic measurements.

Cassé et al. (1975a) have calculated the Cl/Fe ratio for the cases of ^{36}Cl survival and decay. They conclude that a substantial amount of ^{36}Cl has decayed and that $T > 2.7 \times 10^6$ years.

3.2.6. <u>The nuclides that decay only by electron capture and are formed by spallation during propagation</u> can be employed to measure adiabatic deceleration in the solar system (Raisbeck et al., 1973). They can also yield information on possible confinement in source regions (Raisbeck et al., 1975). A detailed discussion of these nuclides is presented by Raisbeck and Yiou in the present volume, in section 6.4. Nuclides of this class are ^{7}Be, ^{37}Ar, ^{41}Ca, ^{49}V, ^{51}Cr, ^{53}Mn, ^{55}Fe, and about 50 more that are heavier than iron.

Limits and constraints on traversal of matter in source regions and on the time scales involved can also be placed by the measurements of γ-ray fluxes from recent supernova remnants and other possible sources of cosmic rays. We shall illustrate this point by considering a specific example: If cosmic rays are produced starting 100 years after a supernova explosion (Schramm et al., 1975), and if we assume (a) that they are then produced over an interval of 10^3 years, (b) that there are 3 supernovae per century in our galaxy (though mostly in obscured regions), and (c) half of the path length of material traversed is in source regions, then about half of the galactic γ-rays should come from about 30 point sources. Future studies of cosmic γ-rays permit tests of assumptions such as presented in the above example.

4. RECOMMENDED STUDIES OF CROSS SECTIONS

In Section 3, we presented an outline of current research on cosmic-ray nuclei. It is obvious that further measurements of cross sections are essential for the progress in this field, especially as isotopic resolution of cosmic-ray nuclides progresses. Large gaps in our knowledge of partial cross sections exist for the abundant elements Ne, Mg and Si. The yield of ^{10}Be from ^{16}O is uncertain. Yields from neutron-rich nuclei like ^{18}O or ^{22}Ne are still nearly unknown. The yields of Ca and Ti from Fe are poorly known. The spallation cross sections of elements at the r- and s- process peaks like Te, Xe, Ba, Pt and Pb need to be studied.

Cross sections of nuclear interactions in air and in detector materials are inadequately known. For example, the spallation cross sections of iron with nitrogen, or with heavier nuclei like those in NaI crystals, or in particle "calorimeters" consisting of tungsten plates, are so inadequately known that semiempirical estimates are not even possible, or in the case of σ (Fe-^{14}N), extremely uncertain. The spallation cross sections of the so-called ultra-heavy cosmic ray nuclei (Z $\geqslant$ 30) in air or in detector materials are completely unknown.

REFERENCES

1.   Bartholomä, K.-P., Siegmon, G. and Enge, W. 1975, ref.
      I.C.R.C.* 1, 343.
2.   Benegas, J.C., Israel, M.H., Klarmann, J. and Maehl, R.C.
      1975a, ref. I.C.R.C. 1, 251.

_____

* This reference, used frequently herein, appears below in its
  sequence; see reference 31.

3. Benegas, J.C., Israel, M.H., Klarmann, J. and Maehl, R.C. 1975b, ref. I.C.R.C. 1, 379.

4. Bertini, H. 1975, Chapter in present volume.

5. Bjarle, C. Herrström, N.-Y., Jacobsson, L., Jönsson, G. and Kristiansson, K. 1975, ref. I.C.R.C. 1, 337.

6. Blake, J.B. and Schramm, D.N. 1974, Ap. Space Sci. 30, 275.

7. Cameron, A.G.W. 1973, Space Sci. Rev. 15, 121.

8. Cartwright, B.G. 1971, Ap. J. 169, 299.

9. Cassé, M. and Soutoul, A. 1974, Symp. Isotopic Comp. of Solar and Galactic Cosmic Rays, Durham, N.H.

10. Cassé, M., Goret, P. and Regnier, S. 1975a, ref. I.C.R.C., 2, 544.

11. Cassé, M., Goret, P. and Cesarsky, C.J. 1975b, ref. I.C.R.C., 2, 646.

12. Chen, K., Fraenkel, Z., Friedlander, G., Grover, Jr., Miller, J.M. and Shimamoto, Y. 1968a, Phys. Rev. 166, 949.

13. Chen, K., Friedlander, G. and Miller, J. M. 1968b, Phys. Rev. 176, 1208.

14. Chen, K., Friedlander, G., Harp, G.D. and Miller, J.M. 1971, Phys. Rev. C4, 2234.

15. Clapham, V.M., Fowler, P.H., O'Ceallaigh, C., O'Sullivan, D. and Thompson, A. 1975, ref. I.C.R.C. 1, 400.

16. Cleghorn, T.F., Freier, P.S. and Waddington, C.J. 1968, Can. J. Phys. 46, S572.

17. Cowsik, R. and Wilson, L.W. 1973, Proc. 13th Int. Cosmic Ray Conf., 1, 500, University of Denver, Denver, Colorado.

18. Cowsik, R. and Wilson, L.W. 1975, ref. I.C.R.C. 2, 659.

19. Cumming, J.B., Haustein, P.E., Stoenner, R.W., Mausner, L. and Naumann, R.A. 1974, Phys. Rev. C10, 739.

20. Dostrovsky, I., Fraenkel, Z. and Friedlander, G. 1959, Phys. Rev. 116, 683.

21. Dwyer, D. and Meyer, P. 1975 ref. I.C.R.C. 1, 390.

22. Fisher, A.J., Hagen, F.A., Maehl, R., Ormes, J.F. and Simon, M. 1975, ref. I.C.R.C. 1, 373.

23. Fisk, L.A., Kozlovsky, B. and Ramaty, R. 1974, Ap. J. 190, L35.

24. Garcia-Munoz, M., Mason, G.M. and Simpson, J.A. 1975a, ref. I.C.R.C. 1, 325.

25. Garcia-Munoz, M., Mason, G.M. and Simpson, J.A. 1975b, I.C.R.C. 1, 331.

26. Geiss, J. 1975, private communication.

27. Ginzburg, V.L. and Ptuskin, V.S. 1975, ref. I.C.R.C. 2, 695.

28. Golden, R.L., Badhwar, G.D. and Stephens, S.A. 1975, ref. I.C.R.C. 2, 672.

29. Hagen, F.A., Fisher, A.J., Ormes, J.F. and Arens, J.F. 1975, ref. I.C.R.C., 1, 361.

30. Havnes, O. 1971, Nature 229, 548.

31. I.C.R.C.—14th International Cosmic Ray Conference, August 1975, Conference Papers, München, Germany, Max-Planck-Institut für Extraterrestrische Physik.

32. Israel, M.H., Price, P.B. and Waddington, J.C. 1975, Physics Today 28, 23.

33. Jacobsson, L., Jönsson, G. and Kristiansson, K. 1975, ref. I.C.R.C. 1, 343.

34. Juliusson, E. 1975 ref. I.C.R.C. 1, 355.

35. Juliusson, E. and Meyer, P. 1975, Ap. J. 201, 76.

36. Karol, P. J. 1974, Phys. Rev. C10, 150.

37. Kristiansson, K. 1971, Ap. Space Sci. 16, 405.

38. Kulsrud, R. M., Ostriker, J. P. and Gunn, J. E. 1972, Phys. Rev. Letters 28, 636.

39. Lagarde-Simonoff, M., Regnier, S., Sauvageon, H., Simonoff, G. N. and Brout, F. 1975, J. Inorg. Nucl. Chem., 37, 627.

40. Lindstrom, P. J., Greiner, G. E., Heckman, H. H., Cork, B. and Bieser, F. S. 1975, to be published in Phys. Rev. Letters.

41. Lund, N., Rasmussen, I. L., Peters, B., Rotenberg, M. and Westergaard, N. J. 1975, ref. I.C.R.C. 1, 263.

42. Maehl, R., Hagen, F. A., Fisher, A. J., Ormes, J. F. and Simon, M. 1975, ref. I.C.R.C. 1, 390.

43. Matsubayashi, T., Noma, M., Saito, T., Sato, Y. and Sugimoto, H. 1975, ref. I.C.R.C. 1, 290.

44. Meneguzzi, M., Cesarsky, C. J. and Meyer, J. P. 1975, ref. I.C.R.C. 2, 652.

45. Mewaldt, R. A., Stone, E. C. and Vogt, R. E. 1975, ref. I.C.R.C. 1, 306.

46. Meyer, J. P. 1975, ref. I.C.R.C. 2, 554.

47. Meyer, J. P. and Cassé, M. 1975, ref. I.C.R.C., late papers volume.

48. O'Dell, F. W., Shapiro, M. M., Silberberg, R. and Tsao, C. H. 1975, ref. I.C.R.C. 2, 526.

49. Ormes, J. F., Fisher, A., Hagen, F., Maehl, R. and Arens, J. F. 1975, ref. I.C.R.C. 1, 245.

50. Orth, C. D., Buffington, A. and Smoot, G. F. 1975, ref. I.C.R.C. 1, 280.

51. Parker, E. N., 1963, Interplanetary Dynamical Processes (Interscience Publishers, Inc., New York), Vol. 8.

52. Perron, C. 1975, Thesis, Orsay.

53. Price, P. B. and Shirk, E. K. 1975, ref. I.C.R.C. 1, 268.

54. Radin, J. R., Gradsztajn, E. and Smith, A. R. 1974, Preprint, submitted to Phys. Rev. C.

55. Raisbeck, G. M., Boerstling, P., Klapisch, R., and Thomas, T. D. 1975, Phys. Rev. C12, 527.

56. Raisbeck, G. M., Comstock, G., Perron, C. and Yiou, F. 1975, ref. I.C.R.C., p. 560.

57. Raisbeck, G. M., Perron, C., Toussaint, J. and Yiou, F. 1973, 13th Int. Cosmic Ray Conf., Denver, Colorado, 1, 534.

58. Raisbeck, G. M. and Yiou, F. 1975, chapter in this volume.

59. Rasmussen, I. L. and Peters, B. 1975, Danish Space Research Institute, preprint.

60. Rudstam, G. 1955, Phil. Mag. 46, 344.
61. Rudstam, G. 1956, Thesis, Uppsala.
62. Rudstam, G. 1966, Z. Naturforsch., 21a, 1027.
63. Schramm, D. N., Hainebach, K. L. and Norman, E. B. ref. I.C.R.C. 2, 448.
64. Shapiro, M. M. and Silberberg, R. 1975a, Phil. Transact. Roy. Soc. (London), A277, 317.
65. Shapiro, M. M. and Silberberg, R. 1975b, ref. I.C.R.C. 2, 538.
66. Shapiro, M. M., Silberberg, R. and Tsao, C. H. 1975, ref. I.C.R.C. 2, 532.
67. Silberberg, R. 1974, Symposium on Isotopic Composition of Solar and Galactic Cosmic Rays, Univ. of New Hampshire, Durham, New Hampshire.
68. Silberberg, R. and Tsao, C. H. 1973a, Ap. J. Suppl. 25, 315, Paper ST-I.
69. Silberberg, R. and Tsao, C. H. 1973b, Ap. J. Suppl. 25, 335, Paper ST-II.
70. Silberberg, R., Shapiro, M. M. and Tsao, C. H., 1975, ref. I.C.R.C. 2, 451.
71. Soutoul, A., Cassé, M. and Juliusson, E. 1975, ref. I.C.R.C. 2, 455.
72. Tsao, C. H., Shapiro, M. M. and Silberberg, R. 1973, Proc. 13th Internat. Cosmic Ray Conf. 1, 107, Denver, Colorado, University of Denver.
73. Webber, W. R., Damle, S. V. and Kish, J. 1972, Ap. Space Sci. 15, 245.
74. Webber, W. R. Lezniak, J. A., Kish, J. and Damle, S.W. 1973a, Astrophys. Space Science, 24, 17.
75. Webber, W. R., Lezniak, J. A. and Kish, J. 1973b, Proc. 13th Internat. Cosmic Ray Conf. 1, 120, Denver, Colo., Univ. of Denver.
76. Yiou, F. and Raisbeck, G. 1972, Phys. Rev. Letters 29, 372.
77. Yiou, F., Raisbeck, G., Perron, C. and Fontes, P. 1973, Conf. Papers, 13th Int. Cosmic Ray Conf., 1, 512.

THE APPLICATION OF NUCLEAR CROSS SECTION MEASUREMENTS TO
SPALLATION REACTIONS IN COSMIC RAYS

G.M. Raisbeck and F. Yiou

Laboratoire René Bernas, du Centre de Spectrométrie
Nucléaire et de Spectrométrie de Masse, B.P. n° 1,
91406 ORSAY, France.

1. INTRODUCTION

Earlier in this meeting we have heard talks on the present
theoretical and experimental situation regarding our understanding
of nuclear spallation reactions*. These studies have been going on
for several decades, essentially ever since the availability of
particle accelerators made it possible to produce such reactions
artificially. As in many areas, however, one finds that what
scientists are able to do in the laboratory, nature has managed to
do long before, often even more efficiently or dramatically. Thus
we find that high energy nuclear reactions associated with cosmic
rays have been going on for millions and probably billions of years
in a number of astrophysical environments. Some of the traces
left by such reactions on these environments are covered in other
presentations at this meeting. What we would like to deal with here
are the effects and implications of nuclear transformations of the
cosmic ray particles themselves, and how those transformations are
simulated in the laboratory. Thus, although the majority of cosmic
rays are protons and alpha particles, it is the small fraction of
heavier species that we will be mainly concerned with here**.
These nuclides interact with the interstellar matter (again mostly
hydrogen and helium) in which they propagate, and thus can undergo
nuclear reactions. For the cosmic ray physicist the effects of
these reactions are both favourable and unfavourable. The unfavourable aspect arises from the fact that the composition of the

---

* The term spallation being used in its broadest sense here to
  refer to any reaction in which the nucleus is broken up.

** See footnote on next page.

*Shen/Merker (eds.), Spallation Nuclear Reactions and Their Applications, 83–111. All Rights Reserved.
Copyright © 1976 by D. Reidel Publishing Company, Dordrecht-Holland.*

cosmic rays is significantly altered, thus tending to mask an
important indication as to their origin. Counterbalancing this is
the fact that the effects left by the nuclear reactions are one of
our most valuable links with the propagation process itself. A
careful unravelling of these effects can thus reveal important
information on where and how this propagation takes place. These
two aspects are discussed further in Sections 2 and 3. In 4 we
consider what type of nuclear information is needed, and in 5 the
techniques that are used to obtain it. In 6 we give some indication
of the present status of the field, using a few examples of current
studies in our own group as illustration. A résumé of measurements
carried out in our laboratory during the past few years is given
in the Appendix.

Although we hope to outline the most important aspects of
this problem, we have not tried to make this a comprehensive
discussion. The treatment is meant to be illustrative rather than
exhaustive. For a similar reason we have given references only
when it seemed necessary to support the text. Several excellent
reviews of this subject, although from a somewhat different standpoint, have been given by Shapiro and Silberberg, the most recent
being in Ref. 2.

2.   SOURCE COMPOSITION

Information on the source composition of cosmic rays can give
us valuable clues as to the potential astrophysical sites for their
formation and subsequent acceleration. For example, early observations that cosmic rays were enriched in heavier ($Z \geqslant 10$) nuclei
compared to " universal " abundances led to speculation that they
came from highly evolved stars. This, together with a favourable
energetic situation, suggested identification of such sources
with supernova. More recently it has become apparent that a very
significant fraction of all heavier element synthesis may occur
in explosive events (3). If this is true, then cosmic rays may
represent one of our most direct samplings of the results of
explosive nucleosynthesis. A knowledge of the " primary " cosmic
ray composition would then contribute not only to a better understanding of their origin, but also to conditions of supernova
nucleosynthesis in general.

** In fact we limit our discussion primarily to species with
$3 \leqslant Z \leqslant 28$. Alpha particles can also suffer spallation to
give ^2H and ^3He. However observation of these isotopes has
only been made at energies where other production processes
and solar modulation complicate the situation considerably. A
rather complete review has been given by Meyer (1). For the
rare cosmic ray nuclides with $Z > 28$ the data is still very
sparce, although spallation is expected to play a significant
role because of large destruction cross sections.

As we shall see below, the quantity of matter encountered by the cosmic rays before reaching the solar system is far from negligible. As a result the arriving abundances of the majority of elements are dominated by (Li, Be, B, F, P, Cl, K, Sc, Ti, V, Cr, Mn) or have very significant contributions of (N, Na, Al, S, Ar, Ca) secondary spallation products. Thus, to obtain the desired information on the primary composition, we need to unfold these nuclear effects, and this requires knowing the amount and distribution of matter encountered, and the relevant nuclear cross sections.

Another aspect of this problem involves the possibility of " selection " effects during acceleration. One extreme point of view for example, is to argue that the cosmic ray source composition is the same as universal composition, and that all the differences we observe are due to these selection effects. A method of testing such ideas is to compare the " enrichment " of two elements differing strongly in the physical property believed to be causing selection. If one wants to test the importance of ionization potential, for example, one might choose to compare Na and Ne. Again it is the abundances before propagation that we are interested in, and in some cases (Na in the present example) the spallation corrections are very significant.

3.   PROPAGATION CONDITIONS

3.1   Quantity and distribution of matter traversed

One of the earliest conclusions arrived at regarding cosmic ray propagation was that primary cosmic rays traverse a significant amount of matter between their acceleration and their arrival in the solar system. This conclusion was based on the ratio Li Be B/C N O, which is believed to be very small universally ($\sim 10^{-6}$) but which is ~ 0.23 in cosmic rays. Assuming a universal ratio at the source, and a knowledge of the relevant cross sections, one can calculate that the primary C N O nuclei must interact with the equivalent of a ~ 4 g/cm^2 slab of hydrogen in order to produce the observed cosmic ray Li Be B. Somewhat later a similar analysis of the spallation products near Fe led several workers to argue that this same " slab " of matter would lead to more heavier secondary products than were actually observed. While the arguement in the latter case was based on estimated cross sections, recent experimental measurements in our laboratory (4) have indeed confirmed that a slab of only ~ 2 g/cm^2 is necessary to give the observed abundances of Sc, V, Cr and Mn.

One way (although not the only one) of avoiding this apparent conflict, is to assume that there is a distribution of potential pathlengths in which the Fe primaries interact with a smaller

amount of matter than the C, O. An example is an exponential
distribution in which, because of their larger destruction cross
sections, the heavy nuclei we see have on average passed through
less material than the lighter ones*. Using this model, and a
mean pathlength of $\sim$ 5 g/cm^2, it appears one can, at least to the
accuracy of present data, thus account for the abundances of all
the largely secondary products in cosmic rays at energies of
$\leq$ 10 GeV/n. It is important to push such comparison as far as
possible since they can help determine whether all cosmic rays
originate in the same sources or whether, for example, Fe nuclei
may be from different sources than C, O. Also, with sufficiently
precise data one could eventually determine the detailed form of
the pathlength distribution, and thus impose certain constraints
on models for the propagation mechanism.

Another question that has been brought to interest recently
is whether cosmic ray primaries of different energies encounter
the same amount of matter. While this was believed true for many
years, recent cosmic ray composition measurements at high (>10 GeV)
energies indicate the opposite (6). These observations show that
the ratio of secondaries/primaries decreases with increasing
energy. This phenomenon has widely been attributed to the fact that
the higher energy nuclei traverse a smaller quantity of matter.
Such an interpretation has very important consequences for cosmic
ray studies. Whether and to what extent this trend continues down
to lower energies is still open to question. It is clear that a
quantitative evaluation of this effect, especially at lower
energies, must correctly unfold any energy variations in the
relevant nuclear cross sections themselves.

3.2   Time and place of propagation

Two of the basic unknowns regarding cosmic ray propagation
are where and for what period of time it takes place. This problem
has tradiationally been considered from the point of view of cosmic
ray " clocks " - i.e. spallation produced radioactive isotopes
having lifetimes of the order of the expected propagation time.
The most important of these species is ^{10}Be, with a rest half-life
of 1.5 x 10^6 years. Other possible candidates are ^{26}Al (7.3 x 10^5
years) and ^{36}Cl (3.0 x 10^5 years). A measure of what fraction of
these radioactive species survives can give us the time between

---

* Such a distribution corresponds to a steady state situation in
  which there is a uniform distribution of sources in the
  galactic disk, with the cosmic rays slowly escaping at the
  boundary regions (the so called " leaky-box " model). With
  some limitations (5) it is also a fairly good approximation to
  more realistic models where containment is due to various types
  of diffusive scattering.

their formation and detection. Clearly, in order to know what
fraction survives, it is necessary to be able to evaluate how
much was produced originally, and this requires knowledge of the
production cross sections.

It is often stated that this temporal information, together
with knowledge of the total amount of matter traversed, can, by
means of the implied density of the propagation medium, tell us
where propagation takes place. This is only strictly true,
however, if production and propagation are contiguous and
homogenous. It is not difficult to imagine situations in which
these assumptions do not hold. For example, production could take
place in a relatively dense medium surrounding the source, followed by propagation in the galactic disk or even in a more tenuous
" halo ". The density inferred from a radioactive " clock " in
this case would not be representative of either the spallation or
the propagation environment. Similarly, as emphasized by Prishchep
and Ptuskin (7), if the diffusion time from the disk to a " halo "
is long compared to the half-life in question, then again the
" radioactive age " will not correctly reflect the propagation
time. One cannot even exclude by this method the possibility of
an extragalactic source, the radioactive age in this case representing the " trapping " time in the galaxy.

In principle, some of these difficulties can be overcome by
determining the age using several isotopes with different half-
lives (or the same isotope at different relativistic energies,
and thus different time dilated half-lives). In practice, such
measurements may be some time away.

Despite these possible ambiguities of interpretation, the
question of cosmic ray " clocks " is a subject which continues to
preoccupy cosmic ray physicists to a considerable degree, and one
in which nuclear cross section measurements play a vital role. As
an example of the problem we show in Fig. 1(a) the predicted
cosmic ray abundances for the isotopes ^{7}Be, ^{9}Be and ^{10}Be, that
we have calculated using current best estimates of source
composition (2), pathlength distribution and nuclear cross sections. It should be realized that such a calculation involves
several hundred cross sections from many progenitor nuclides at
many energies. Fortunately in this particular case the most
important of these are measured experimentally, and the results
reasonably insensitive to uncertainties in the unmeasured ones.

One immediately sees two things in Fig. 1(a). First, the
abundance of ^{10}Be compared to all Be is very small. Thus any
conclusions based on Be elemental abundance are, at best, very
delicate. Secondly, although the absolute production rates vary
significantly with energy, the relative rates are much more

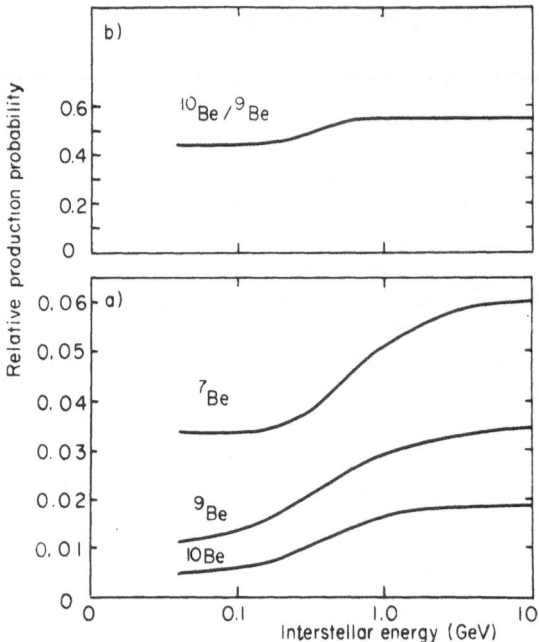

Fig. 1 - Calculated abundance of spallation produced Be isotopes
in cosmic rays as a function of their interstellar energy,
assuming complete survival of ^{7}Be and ^{10}Be. Calculation done
assuming 5 g/cm^2 exponential pathlength, power law source spectra
in total energy E$^{-2.6}$, and source composition from Ref. 2.
(a) Predicted abundances relative to ^{12}C
(b) Predicted ^{9}Be/^{10}Be ratio.

constant. This is seen in Fig. 1(b) where the ratio ^{10}Be/^{9}Be is
plotted. At low (< 1 GeV) energies this relative energy independence has a very important practical advantage because, due to
solar modulation effects, the cosmic ray detection energy differs
(by an uncertain amount) from the actual interstellar energy.
However, as seen in Fig. 1(b), the " non-decay " ratio of ^{10}Be/^{9}Be
will be fairly insensitive to these modulation effects. This is
only one of the reasons we have tried to emphasize the importance
of making actual isotopic measurements of cosmic ray ^{10}Be in order
to resolve the " age " question (8).

4.   TYPE OF NUCLEAR DATA NEEDED

    It is evident from the above discussion that the information
needed regarding nuclear reactions are the total cross sections
leading to various spallation products. Thus, detailed theoretical
or experimental information regarding reaction mechanisms is
irrelevant (except, of course, to the extent that it permits a
more reliable or systematic estimate of unmeasured cross sections).

Also, because of the astrophysical time scale involved, the data we are interested in are the cumulative cross sections of stable or very long lived products. Thus convential radiochemical techniques, which are traditionally the most common method of determining total cross sections, are only of limited value. It is probably because of the above two factors that this area of research has received the attention of very few workers, despite, as outlined by J. Hudis at this meeting, the relatively large number of studies devoted to spallation type reactions in general.

The decision as to what specific cross sections are critical is naturally dependent on which of the previously discussed areas one is interested in. For example, if one is trying to evaluate the quantity of matter traversed then one looks for a product nuclide which is expected to have very low abundance at the source (e.g. Li Be B, F, Sc). If one is looking at the question of cosmic ray " age " then one focuses on the favourable radioactive " clocks" (^{10}Be, ^{26}Al, ^{36}Cl). For the question of source composition the most important products are those elements where spallation is an important, but not overwhelming, contributor to the observed abundance.

The targets* that are in general most important are those species (C, O, Ne, Mg, Si and Fe) that have large source abundances. Unfortunately those doing spallation studies are usually unaware of this and choose their target material because it is easy to work with or readily available. Thus, there are dozens of papers in the literature dealing with Al and Cu spallation, and relatively few for Si and Fe.

The energy region of interest is, in principle, all the way from the spallation threshold up to the highest energies observed in cosmic rays. In reality, for the present purposes we are mainly concerned with the range between $\sim$ 100 MeV and 10 GeV. According to current modulation theory, cosmic ray nuclei of less than $\sim$ 100 MeV are virtually excluded from the earth's orbit[†]. Above $\sim$ 10 GeV nuclear cross sections appear to reach a constant value.

---

*The term target here refers to the cosmic ray progenitor nuclide. In most studies this "projectile" ion is simulated by a stationary "target" nucleus in the laboratory system, and an energetic proton (alpha) represents the interstellar H(He). At the same relative velocity (energy/nucleon) the nuclear cross sections are, of course, identical. The use of relativistic heavy ions is discussed in Section 5.6.

†The importance of lower energy cross sections for the nucleosynthesis of the light elements in cosmic rays, treated elsewhere at this meeting, may on the other hand be very significant.

We should perhaps also mention here that it is isotopic cross sections that are, and will become increasingly so, important to the cosmic ray problem. This is because the cosmic ray source and spallation produced isotopic distributions are believed to be very different. Thus, when as expected, cosmic ray isotopic measurements become available in the near future, the unfolding and utilization of spallation effects will become a much more sensitive procedure.

5. EXPERIMENTAL TECHNIQUES

The requirements outlined in the preceeding section impose certain constraints on the experimental techniques applicable to cross section measurements relevant to the cosmic ray question. We describe very briefly below several techniques which have been applied to this problem. The reader is referred to the appropriate references for further details.

5.1 Mass spectrometry

The most successful technique utilized to date has been that of mass spectrometry. In this procedure, as applied in the Orsay laboratory (4, 9-15), the target material of interest is first irradiated with an intense flux of high energy protons (or alpha particles). The small quantity (10^{-9} - 10^{-12} g) of product nuclei is then extracted from the target, and the relative isotopic abundances measured on a high sensitivity sputtering mass spectrometer. The absolute cross sections are obtained by using isotope dilution procedures or with reference to a known radioactive cross section. This method has the advantage of being applicable at all energies and, in principle, to all combinations of target and product. Its chief disadvantage is that it is very painstaking and requires virtually a new procedure for each target-product combination.

5.2 Time-of-flight

A very elegant technique, that so far has only been applied at relatively low energies (16-20), involves bombarding a thin target with protons and measuring the velocity and energy of the emitted nuclei. Such information is, of course, only sufficient to identify the mass of the product. However, in many cases where there is only a single stable isotope, it is this isobaric yield which is relevant to the propagation problem. The potential of this technique is therefore considerable.

5.3   dE/dx-E

The identification of reaction products by means of their
rate of energy loss (dE/dx) in solid state detectors has been a
very powerful tool in a variety of nuclear physics applications.
In order to measure total cross sections by this method it is
necessary to record fragments over most of their energy spectrum.
This is reasonably easy when the target is heavy, and the energies
of the emitted fragments large (21, 22), but more difficult for
the lighter targets relevant to the cosmic ray question. Never-
theless, by using a combination of thin detectors, this method
has been used to measure Li Be and B cross sections from a Ni
target (23). As was pointed out in that work, the possibility of
applying this technique to even lighter targets is limited by the
resolution obtainable with the thinner detectors necessary to
observe lower energy events. Thin gas counters as the dE/dx
element do permit lower energies to be reached, but usually only
resolve elements. One possibility of overcoming this is by
combining a gas counter with the time-of-flight technique mentioned
above to get both Z and A information down to very low fragment
energies.

5.4   Nuclear emulsion

Attempts have been made to apply this technique (24, 25), but
the results have in general been in disagreement with better
established procedures. As well as having the difficulty of
identifying the target species, there are certain modes of breakup
(involving more than one neutron) that cannot be uniquely
determined by kinematics.

5.5   Radioactivity

There are two exceptions to the general statement made
earlier about the inapplicability of radiochemical measurements
to the cosmic ray spallation problem*. First, there is a certain
group of nuclides which decay by pure electron capture under
laboratory conditions, but in a cosmic ray environment will be
stable (see section 6.4). Measurement of these cross sections
(e.g. ^{7}Be, ^{51}Cr) can thus be carried out using conventional
radiochemical methods. The other exception are those radionuclides
having half-lives of the order of the cosmic ray " age ", and
being important for precisely this reason, as discussed in

---

* In addition one should not forget the importance of radio-
   chemically determined cross sections used in " normalizing " the
   results of several of the other techniques discussed here.

section 3.2. These cross sections can also be determined by
radioactivity, although their very low activities usually require
somewhat specialized procedures. For example, we have determined
the cross section of ^{10}Be in ^{11}B, N, Mg and Si using an isotope
separator to isolate the ^{10}Be from other activities (26, 27).
Other workers have determined ^{26}Al and ^{36}Cl using extensive
chemical separation procedures and low level γ or β counting
(28, 29).

5.6 Relativistic heavy ions

    It would be inappropriate to finish this section without
noting a very recent, but very powerful technique, which promises
to have a significant impact on many of the problems discussed
here. This is the use of magnetic rigidity and energy loss
measurements to identify products emitted in the spallation of
relativistic heavy ions (30-32). In such a system, the role of the
progenitor nucleus and hydrogen target can be as in the cosmic
rays themselves. The result is that the product species, which
leave the target with essentially the same velocity (energy/nucleon)
as the initial projectile, are much more amenable to physical
measurement. The great advantage of this technique is its generality.
Thus, using the same experimental arrangement, it should be
possible to study any target-product combination, once appropriate
beams are available.

    There are other advantages of this technique which are
relevant to the cosmic ray question. For example, it will permit
the study of spallation of rare isotopes or even, through the use
of secondary beams, radioactive nuclides. It will also permit the
simulation of cosmic ray breakup in the atmosphere or in detector
materials. While not directly relevant to interstellar propagation,
corrections for such interactions are still a major source of
uncertainty in all balloon measurements of cosmic ray abundances.

    The limitations of the technique appear to be in the energy
range that can be covered. At low energies the finite distribution
of transverse momenta may become significant compared to the total
fragment momentum, thus making it more difficult to separate
adjoining isotopes. The upper energy limit is imposed by the
maximum beam energy (2.5 GeV/n for A/Z = 2) that can be attained
by the Bevalac facility, which is presently the only accelerator
where such beams are available. Such an energy is still somewhat
below the asymptotic region of constant cross sections for
progenitors like Fe.

    The first results using this technique have recently become
available (32), and some of the cross sections relevant to the
propagation process are shown in Table I. We have included in

TABLE I

A comparison of Li Be B production cross sections from ^{12}C and ^{16}O as determined by mass spectrometric and relativistic heavy ion techniques

^{12}C (p, x)

Energy (GeV) Product	0.6 (a)	1.05 (b)	2.1 (b)	25 (a)
^{6}Li	15.5 ± 2.2*	14.8 ± 1.5	12.44 ± 2.2	15.5 ± 3
^{7}Li	13.6 ± 2.1	11.0 ± 1.0	10.4 ± 0.8	14.5 ± 3
^{7}Be	11.0 ± 1.1	9.49 ± 0.99	8.45 ± 0.81	9.1 ± 0.4
^{9}Be	5.3 ± 0.8	5.92 ± 0.54	5.13 ± 0.54	6.1 ± 0.9
^{10}Be	2.8 ± 0.5	3.42 ± 0.35	3.41 ± 0.35	3.6 ± 0.6

^{16}O (p, x)

	0.6 (a)	2.1 (b)	19 (a)
^{6}Li	12.4 ± 2.4	13.9 ± 2.4	14.4 ± 2.7
^{7}Li	11.3 ± 2.2	11.1 ± 1.3	14.2 ± 2.6
^{7}Be	7.0 ± 1.7	10.1 ± 1.2	10.8 ± 1.4
^{9}Be	2.6 ± 0.94	4.17 ± 0.55	3.6 ± 1
^{10}Be	0.6 ± 0.24	2.05 ± 0.31	1.0 ± .4
^{10}B	12 ± 5	10.69 ± 1.8	15 ± 8
^{11}B	25 ± 12	26.4 ± 2.6	< 45

* All cross sections are in mb and are cumulative (i.e. ^{6}Li + ^{6}He, ^{11}C + ^{11}B, ^{10}C + ^{10}B)

(a) Mass spectrometric results (see Appendix for refs)

(b) Heavy ion results (ref. 32)

this table the earlier mass spectrometric results from our own
laboratory for comparison. One sees that, while the Berkeley
measurements are given with a smaller uncertainty, the agreement
is in general remarkably good, and in most cases well within the
quoted errors. There is one exception, and that is for ^{10}Be in
oxygen, where the discrepancy is about a factor of 2. Because of
the importance of ^{10}Be, this disagreement is, of course, rather
important, and bears further investigation. Over all, however, we
would say the general agreement is quite satisfactory, and should
be a source of gratification to both experimental groups (as well
as considerable comfort to the cosmic ray workers who must use
such measurements).

5.7 Semi-empirical formulas

While certainly not an experimental technique, it may be
appropriate at this point to say a few words about methods of
estimating cross sections which have not been measured. Several
years ago the necessity for including a large number of unmeasured
cross sections in the propagation problem motivated Silberberg
and Tsao (33) to formulate a series of semi-empirical relationships,
based on an earlier idea of Rudstam (34). Unlike most of the models
discussed by Bertini at this meeting, the goal of this formula is
not to try to understand the physics of the spallation reactions,
but rather simply to empirically fit the existing experimental
data as well as possible. As with all such formulations, the
reliability of the predicted values is thus dependent on the
extent and accuracy of the data on which it is based. It is clear,
for example, that the most accurate predictions are expected for
types of reactions for which there are already fairly complete
data. It is also clear that such a formula is most useful when a
large number of its predictions are used together since then the
errors tend to compensate each other (this is not to say that for
a given set of similar reactions there may not be systematic
inaccuracies due to insufficient or even incorrect input data).
In any event, while there may be some difference of opinion as to
the exact uncertitude to assign to the predictions of this formula,
it is evident that the sheer magnitude of the complete cosmic ray
transport equation ensures that such a device will continue to
serve a very useful purpose in estimating otherwise unavailable
cross sections.

6. PRESENT STATUS AND SOME CURRENT STUDIES

In this last section we try to indicate what the present status
is regarding the interplay between cosmic ray physics and spallation studies, and what directions these studies can be expected
to take in the future. That there is an interplay, of course,
implies that progress is dependent not only on the nuclear

information, but also on cosmic ray observations. In particular, as noted in Section 4, much of the nuclear data that is available today will be fully exploited only when the cosmic ray isotopic composition has been measured. Conversely, interpretation of a number of cosmic ray observations is presently limited by the lack of adequate cross section information.

Regarding the source composition we have reliable estimates really only for those elements with large primary abundances (C, O, Ne, Mg, Si, Fe) where spallation contributions do not play a significant role. A recent and illustrative example of this is the substantial revision in the calculated source N abundance (35). This revision, which is due largely to the measurement of the production cross sections of ^{14}N and ^{15}N in ^{16}O, by Lindstrom et al. (32), has a number of interesting implications regarding source conditions.

As far as our information on the propagation process itself, the bulk of the attention to date has revolved around the light elements Li, Be and B. This is natural since, historically they were the first to be identified as spallation products and, because of their abundances, fairly accurate cosmic ray observations were possible. Today it seems fair to say that we believe we understand this situation reasonably well. A fairly complete and reliable set of experimental cross sections is available. The elemental and, to the extent they are available, the isotopic abundances of these elements in cosmic rays appear to be in quantitative accord with what we expect from a pure spallation origin. The outstanding problem yet to be resolved is the presence or absence of ^{10}Be, and here the onus seems to be mainly on the cosmic ray observer.

As cosmic ray measurements have become more extensive and accurate there has been, for the reasons given above, an increased necessity to expand the scope of the spallation studies also. We choose to illustrate this by discussing briefly four examples of studies which have recently been undertaken in our laboratory. Such a choice is naturally very personal, and is not meant to imply any value judgement on the importance of other questions.

6.1 Spallation of Fe

A comparison of the cosmic ray abundances of the elements directly below Fe in the periodic table (Sc, V, Ti, Cr, Mn) with their generally accepted universal (i.e. solar system) abundances, gives an " overabundance " in cosmic rays by 1 to 2 orders of magnitude above and beyond the Fe overabundance itself (2). Thus, like Li Be B, these species too are believed to be formed prima-rily of spallation products. As mentioned in section 3.1, even a

rather crude consideration of these products helped to establish
the idea of an exponential (or similar) pathlength distribution.
Recent improvements in the cosmic ray measurements now make it
possible to consider these elements on a more quantitative basis.
On the nuclear side, a very important contribution to this cause
is the recent work of Perron (4) who has measured cross sections
for the production of the stable isotopes of these elements at
600 MeV and 21 GeV. These measurements (see Appendix) and propagation calculations carried out using their results, show that the
observed cosmic ray abundances can indeed be accounted for by the
spallation process. Such calculations cannot at present, however,
exclude small source contributions of the order of their universal
abundances. As explained earlier, knowledge of the cosmic ray
isotopic composition would permit even more stringent limits to
be set. The determination of the source abundances in the Fe
region is particularly important because it is these species that
are believed to be formed during the explosive nucleosynthesis,
and are most sensitive to the parameters in such an event. An
extension of these studies to include elements such as Ca, where
the source contributions are comparable to the spallation, is
thus highly desirable.

   Other important reasons for studying the spallation products
of Fe are that they include several potential " clock " nuclei
(^{36}Cl, ^{54}Mn), and a very fascinating group of nuclei that decay
by pure electron capture (see Section 6.4).

6.2 Alpha induced reactions

   Since the interstellar medium is believed to be $\sim$ 10 % helium,
alpha induced spallation reactions contribute to the observed
products in a non-negligible way. In most propagation calculations
these interactions have usually been ignored, or grossly approximated. The main reason for this is the almost complete absence
of experimental data in the relevant energy region. This, in turn,
was because, until recently, there were no accelerators providing
alpha beams with energies greater than 230 MeV/n. This situation
has changed, and there are now several accelerators offering
alpha beams of > 1 GeV/n. A program has therefore begun in our
laboratory to investigate such reactions using the " Saturne "
synchrotron at Saclay (14, 36). The question we are interested
in is how does the ratio σ_α/σ_p (alpha/proton cross sections)
vary (a) for different reactions at the same energy/nucleon (the
relevant parameter for the propagation problem) ; (b) as a function of energy for a given reaction. It is these factors that
will determine how, and to what extent, inclusion of the interstellar helium will modify results compared to a pure hydrogen
interstellar medium. Some preliminary answers to these questions
are seen in Table II, where we have compared some alpha cross

TABLE II

A comparison of some alpha and proton induced cross sections

(from Ref. 36)

Target	Product	$\dfrac{\sigma_\alpha\,(2.8\ \mathrm{GeV})}{\sigma_p\,(0.6\ \mathrm{GeV})}$	$\dfrac{\sigma_\alpha\,(4.6\ \mathrm{GeV})}{\sigma_p\,(1\ \mathrm{GeV})}$	$\dfrac{\sigma_\alpha\,(4.6\ \mathrm{GeV})}{\sigma_p\,(23\ \mathrm{GeV})}$
C	^{7}Be	1.48	1.93	1.77
	^{9}Be	1.89	1.95	1.64
	^{10}Be	1.92	1.58	1.49
O	^{7}Be	2.33	1.73	1.51
Mg	^{7}Be	2.57	2.08	1.79
	^{22}Na	1.04	1.28	1.50
Si	^{7}Be	3.19	2.51	1.79
	^{22}Na	1.06	1.18	1.50
Fe	^{7}Be	8.16	5.08	1.74
	^{22}Na	11.7	3.97	1.59
	^{46}Sc	1.12	1.03	1.44
	^{48}V	0.92	1.01	1.53
	^{51}Cr	0.93	1.08	1.45
	^{52}Mn	1.00	1.23	1.65
	^{54}Mn	1.35	1.25	1.43

sections at 2.8 and 4.6 GeV with proton measurements at approximately the same energy/nucleon. It is evident that σ_α/σ_p is not constant, either as function of velocity, nor for different reactions at a given velocity. On the other hand, at sufficiently high energies (right hand column of Table II) this ratio appears to be relatively constant at about 1.4 - 1.8. The implications of these results to the cosmic ray problem are discussed in more detail elsewhere (36). A number of the cross sections are given in the Appendix.

6.3 Energy variations

At the epoch when measurements seemed to indicate the composition of cosmic rays was independent of energy, little attention
was given to the variation of nuclear cross sections with energy.
However, now that variations are apparent, and particularly
variations in the secondary/primary ratios, it has become important
to identify, in a fairly quantitative manner, exactly what the
nuclear excitation functions look like. At the Laboratoire René
Bernas we are therefore in the process of measuring a number of
these excitation functions in the energy range 0.6 to 25 GeV (37,
38). For analogous motives, we have also extended measurements
of Li and Be cross sections in a carbon target up to 300 GeV (15).
Some results are included in the Appendix. The overall conclusion
of these studies is that there are certain variations in cross
sections at energies < 10 GeV which can be important to the
cosmic ray problem, while above this energy the cross sections
are essentially constant.

6.4 Electron capture isotopes

There is one group of cosmic ray spallation products whose
behavior is peculiar to their astrophysical environment. These
are isotopes that normally decay by orbital electron capture. In
the absence of such electrons, these nuclides will of course be
perfectly stable. Indeed, at the energies characteristic of
cosmic rays, it is normal to think of these ions as fully
" stripped ". In fact there is a certain probability that such a
high energy ion can pick up an electron during its propagation,
and this probability depends strongly on its energy. If one works
out the problem quantitatively, it turns out that above a certain
characteristic energy ($\sim$ 500 MeV/n for ^{49}V, $\sim$ 25 MeV/n for ^{7}Be),
these nuclides are stable, while below this energy they have a
rapidly varying probability for decay (39, 40). In addition, this
decay probability is dependent on the density of the medium in
which they are propagating (at high densities the electron can
be restripped off). This behavior makes these isotopes potentially
important " probes " of their propagation environment (40). For
example, they may be able to give more direct information on the
propagation density than the radioactive " clocks " discussed in
Section 3.2. Also they can give information on the solar modulation process since their survival is a rapidly dependent function
of their interstellar energy, which is different than their detection energy at the earth's orbit (41).

In order to take advantage of these properties it is necessary
to know both the survival abundance and the production rate of
the isotope in question. The best way to do this is to consider
the relative abundance of the electron capture isotope and a

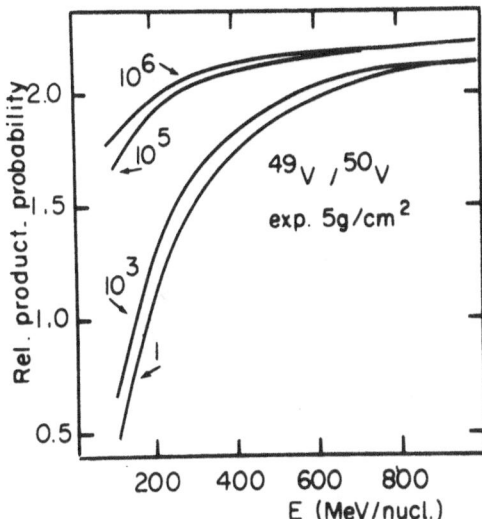

Fig. 2 - Calculated ratio of spallation produced $^{49}V/^{50}V$ in
cosmic rays as function of density of propagation medium
(in atoms/cm^3) and interstellar energy. Calculation assumes
exponential pathlength of 5 g/cm^2 and power law source spectrum
in total energy, $E^{-2.6}$.

neighbouring stable isotope. We show, for example, in Fig. 2,
the predicted ratio $^{49}V/^{50}V$ in cosmic rays as a function of
interstellar energy, and for different assumed densities of the
propagation medium. In Fig. 3 the same ratio is considered, this
time as a function of energy at the earth's orbit, assuming a
nominal interstellar density of 1 atom/cm^3. The calculation of
the modulation effect has been carried out using the model of
Gleeson and Urch (42, 43), with parameters applicable to a solar
minimum (1965) and solar maximum (1970).

If one considers that there are a number of these isotope
ratios, with their survival ratios depending in different ways
on the parameters used in the calculations, one gets some idea
of the potential wealth of information that they contain. It is
also clear that rather precise cross section information is
needed here, since a 50 % error in the predicted ratio of Fig. 2
or 3, for example, could lead to quite erroneous conclusions
regarding the survival probability. Thus, this is a case where
approximate cross section estimates, such as those from semiempirical formulas, are not very satisfactory. In fact, it was

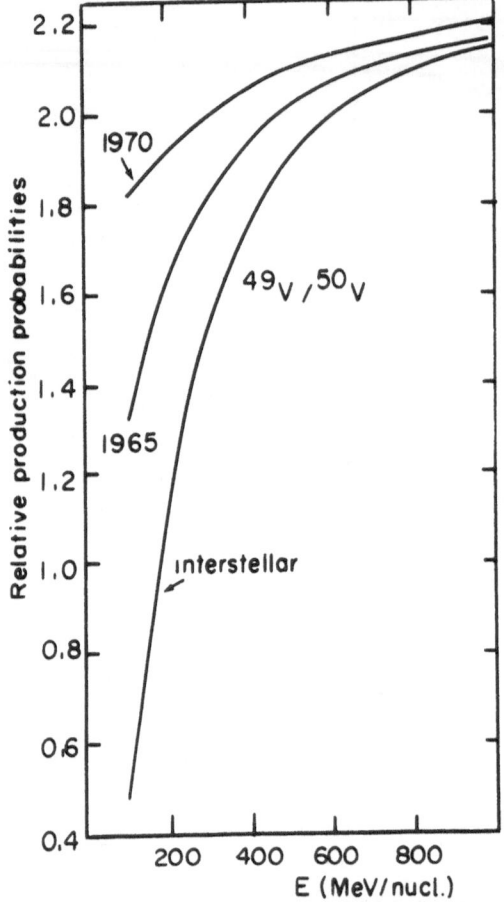

Fig. 3 – Calculated cosmic ray $^{49}V/^{50}V$ ratio as modulated at earth's orbit for 1965 and 1970. Interstellar input data are the same as Fig. 2 for density of 1 atom/cm^3 and modulation parameters are from Ref. (42, 43).

our interest in having accurate experimental data regarding the behavior of these isotopes which was the primary motivation for initiating the Fe spallation studies described in Section 6.1.

ACKNOWLEDGEMENTS

We would like to acknowledge the efforts of all those colleagues at the Laboratoire René Bernas, past and present, who have contributed to the measurements discussed here. Heading such a list must, of course, be René Bernas himself. His foresight

in recognizing the importance of these studies, and his enthu-
siasm for pursuing them, have been a significant factor in
enabling such a program to continue after his untimely passing
away.

Parts of this review have been taken from an invited paper
given by one of the authors (GMR) at the 2nd High Energy Heavy
Ion Summer Study, Lawrence Berkeley Laboratory, Berkeley, California, July 1974.

REFERENCES

(1) J.P. Meyer, Thesis, University of Paris-Sud, Orsay (1974).

(2) M.M. Shapiro and R. Silberberg, Phil. Trans. Royal Society,
      (London), A $\underline{277}$, 319 (1975).

(3) See for example " Explosive Nucleosynthesis ", Ed. by D.N.
      Schramm and W.D. Arnett, University of Texas Press (1973).

(4) C. Perron, Thesis, University of Paris-Sud, Orsay (1975).

(5) V.S. Ptuskin, Astrophys. and Space Sci., $\underline{28}$, 17 (1974).

(6) For a review see : W.R. Webber, Rapporteur paper at 13th Int.
      Cosmic Ray Conf., Denver, Colo. (1973).

(7) V.L. Prishchep and V.S. Ptuskin, Astrophys. and Space Sci.,
      $\underline{32}$, 265 (1975).

(8) G.M. Raisbeck and F. Yiou, Proceeding of the 13th Int. Cosmic
      Ray Conf., Denver, Colo. (1973) Vol. 1, p. 494.

(9) F. Yiou, Ann. Phys. (Paris), $\underline{3}$, 169 (1968).

(10) P. Fontes, C. Perron, J. Lestringuez, F. Yiou and R. Bernas,
      Nucl. Phys., A $\underline{165}$, 405 (1971).

(11) G.M. Raisbeck, J. Lestringuez and F. Yiou, Phys. Rev., C $\underline{6}$,
      685 (1972).

(12) G.M. Raisbeck, J. Lestringuez and F. Yiou, Nature Phys. Sci.,
      $\underline{244}$, 28 (1973).

(13) F. Yiou, P. Fontes, C. Perron and G.M. Raisbeck, Proceedings
      of the 13th Int. Cosmic Ray Conference, Denver, Colo.,
      (1973) Vol. 1, p. 512.

(14) G.M. Raisbeck and F. Yiou, Phys. Rev. Lett., $\underline{35}$, 155 (1975).

(15) G.M. Raisbeck, J. Lestringuez and F. Yiou, Phys. Lett.,
      $\underline{57}$ B, 186 (1975).

(16) C.N. Davids, H. Laumer and S.M. Austin, Phys. Rev., C $\underline{1}$,
      270 (1970).

(17) H. Laumer, S.M. Austin, L.M. Panggabean and C.N. Davids,
      Phys. Rev., C $\underline{8}$, 483 (1973).

(18) H. Laumer, S.M. Austin and L.M. Panggabean, Phys. Rev., C 10, 1045 (1974)

(19) D.L. Oberg, D. Bodansky, D. Chamberlin and W.W. Jacobs, Phys. Rev., C 11, 410 (1975).

(20) C. Roche, Ph.D. Thesis, University of Maryland, (1974), (unpublished).

(21) A.M. Poskanzer, G.W. Butler and E.K. Hyde, Phys. Rev., C 3, 882 (1971).

(22) E.K. Hyde, G.W. Butler and A.M. Poskanzer, Phys. Rev., C 4, 1759 (1971).

(23) G.M. Raisbeck, P. Boerstling, R. Klapisch and T.D. Thomas, Phys. Rev., C 12, 527 (1975).

(24) M. Jung, C. Jacquot, C. Baixeras-Aiguabella, R. Schmitt and H. Braun, Phys. Rev., C 1, 435 (1970).

(25) M. Jung, C. Jacquot, C. Baixeras-Aiguabella, R. Schmitt, H. Braun and L. Girardin, Phys. Rev., 188, 1517 (1969).

(26) G.M. Raisbeck and F. Yiou, Phys. Rev. Lett., 27, 875 (1971).

(27) G.M. Raisbeck and F. Yiou, Phys. Rev., C 9, 1385 (1974).

(28) M. Honda and D. Lal, Phys. Rev., 118, 1618 (1960).

(29) S. Regnier, M. Lagarde, G.N. Simonoff and Y. Yokoyama, Earth and Planetary Sci. Lett., 18, 9 (1973).

(30) H.H. Heckman, D.E. Greiner, P.J. Lindstrom and F.S. Bieser, Phys. Rev. Lett., 28, 962 (1972).

(31) D.E. Greiner, P.J. Lindstrom, H.H. Heckman, B. Cork and F.S. Bieser, Phys. Rev. Lett., 35, 152 (1975).

(32) P.J. Lindstrom, D.E. Greiner, H.H. Heckman, B. Cork and F.S. Bieser, Lawerence Berkeley Laboratory report LBL-3650 (1975) (unpublished).

(33) R. Silberberg and C.H. Tsao, Ap. J. Supp. n° 220, 25, 315 (1973).

(34) G. Rudstam, Z. Naturf., 21 A, 1027 (1966).

(35) M.M. Shapiro, R. Silberberg and C.H. Tsao, Bull. Am. Phys. Soc., 20, 710 (1975).

(36) G.M. Raisbeck and F. Yiou, Presented at the 14th Int. Cosmic Ray Conf., Munich, Germany (1975), paper OG 9.2-3.

(37) G.M. Raisbeck and F. Yiou, Presented at the 14th Int. Cosmic Ray Conf., Munich, Germany (1975), paper OG 9.2-1.

(38) G.M. Raisbeck and F. Yiou, Phys. Rev., C 12, (in press).

(39) F. Yiou and G.M. Raisbeck, Astrophys. Lett., 7, 129 (1970).

(40) G.M. Raisbeck, G. Comstock, C. Perron and F. Yiou, Presented at the 14th Int. Cosmic Ray Conf., Munich, Germany (1975), paper OG 9.3-8.

(41) G.M. Raisbeck, G. Comstock, C. Perron and F. Yiou, Presented at the 14th Int. Cosmic Ray Conf., Munich, Germany (1975), paper MG 2.9.

(42) G.L. Gleeson and I.H. Urch, Astrophys. and Space Sci., 11, 288 (1971).

(43) I.H. Urch and G.L. Gleeson, Astrophys. and Space Sci., 17, 426 (1972).

(44) J. Audouze, M. Epherre and H. Reeves in " High Energy Nuclear Reactions in Astrophysics ", Ed. by B.S.P. Shen, W.A. Benjamin Inc., New York (1967).

(45) D.V. Reames in " High Energy Nuclear Reactions in Astrophysics", Ed. by B.S.P. Shen, W.A. Benjamin Inc., New York (1967).

(46) J.P. Cumming, Ann. Rev. Nucl. Sci., 13, 261 (1963).

(47) J.R. Radin, E. Gradsztajn and A.R. Smith, Lawrence Berkeley Laboratory report LBL-2680 (1975), (to be published).

(48) J. Radin, A. Smith and N. Little, Phys. Rev., C 9, 1781 (1974).

(49) J. Lestringuez, G.M. Raisbeck, F. Yiou and R. Bernas, Phys. Lett., 36 B, 331 (1971).

(50) F. Yiou and G.M. Raisbeck, Phys. Rev. Lett., 29, 372 (1972).

(51) C. Orth, H.A. O'Brien, M.E. Schillaci, B.J. Dropesky, J. Cline, E.B. Nieschmidt and R.L. Brodzinski (preprint).

APPENDIX

In a previous volume of this type, a survey was given of
existing experimental cross section data relevant to astrophysical spallation processes (44), and another of outstanding
measurements that were still needed for the cosmic ray propagation
question (45). Since that time a considerable amount of work has
been carried out at the Laboratoire René Bernas in this field.
These measurements are to be found scattered in a variety of
sources, and we have often been asked if there existed an
up-to-date compilation of that work. We therefore felt that it
might be useful to present such a summary here. The résumé is
complete up to August 1975, including a number of recent unpublished or " in-press " results that are discussed in the text.
For the reasons indicated in the text we have restricted the
coverage to energies of > 100 MeV and stable or long lived
products, including pure electron capture isotopes (see Section
6.4). In cases where earlier data have been modified or superseded
by later work, we give only the most recent results.

We would like to urge those making use of the data to consult,
and cite, the original sources. This is not only fair to the
individual workers involved, but is the only means by which one
can make a reliable assessment of how the quoted uncertainties
were arrived at, and what they include. Such an assessment is
often very critical to the problems discussed in the text (8).

Finally, while the results listed below fall considerably
short of satisfying all the requirements cited in Ref. (45), it
is clear that significant progress has been made since that time.
It is hoped that with the increased awarness of this problem,
and the new techniques being brought to bear on it, that this
progress will continue at an even greater rate in the future.

APPENDIX   TABLE   I

Some proton induced cross sections relevant to cosmic ray spallation

Target	Product	Energy (GeV)	σ^{*} (mb)	Refs	Notes
^{10}B	^{7}Be	.15	8.0 ± 1.6	26	
		.60	5.4 ± 1.1	26	
^{11}B	^{7}Be	.15	6.9 ± 1.4	26	
		.60	4.2 ± 0.8	26	
	^{10}Be	.15	11 ± 4	26	a
		.60	14 ± 5	26	a
^{12}C	^{6}Li	.15	11.6 ± 1.6	11	
		.60	15.5 ± 2.2	11	
		25	15.3 ± 3.0	13	
		300	14.1 ± 1.8	15	
	^{7}Li	.15	9.2 ± 1.6	11	
		.60	13.6 ± 2.1	11	
		25	14.5 ± 3.0	13	
		300	12.7 ± 2.0	15	
	^{7}Be	.15	12.1 ± 1.2	46	b
		.60	11.0 ± 1.1	46	b
		25	9.1 ± 0.4	46	b
		300	9.1	15	c
	^{9}Be	.15	3.2 ± 0.5	10	d
		.60	5.3 ± 0.8	10	d
		25	6.1 ± 0.9	13	
		300	6.1 ± 0.9	15	
	^{10}Be	.15	1.1 ± 0.2	10	d
		.60	2.8 ± 0.5	10	d
		25	3.6 ± 0.6	13	
		300	3.7 ± 0.7	15	

Target	Product	Energy (GeV)	σ * (mb)	Refs	Notes
^{13}C	^{7}Be	.15	6.7 ± 1.2	12	
		.60	4.9 ± 0.9	12	
		25	3.7 ± 1.0	13	
	^{9}Be	.15	7.0 ± 1.5	12	
		.60	7.2 ± 1.5	12	
		25	6.4 ± 0.7	13	
	^{10}Be	.15	4.0 ± 0.9	12	
		.60	6.0 ± 1.4	12	
		25	5.9 ± 1.7	13	
^{14}N	^{10}Be	.15	0.6 ± 0.2	27	
		.60	1.8 ± 0.6	27	e
^{16}O	^{6}Li	.135	10.0 ± 1.8	9	
		.60	12.4 ± 2.4	9	
		19	14.4 ± 2.6	9	
	^{7}Li	.135	8.5 ± 1.7	9	
		.60	11.3 ± 2.2	9	
		19	14.2 ± 2.7	9	
	^{7}Be	.135	5.4 ± 1.0	9	
		.60	7.0 ± 1.7	9	
		1.0	9.4 ± 1.5	37	
		2.0	9.8 ± 1.4	37	
		19	10.8 ± 1.4	9	
	^{9}Be	.135	1.7 ± 0.4	9	
		.60	2.6 ± 1.0	9	
		19	3.6 ± 1.0	9	
	^{10}Be	.135	.37 ± .12	9	
		.60	0.6 ± .24	9	
		19	1.0 ± 0.4	9	

Target	Product	Energy (GeV)	σ * (mb)	Refs	Notes
^{16}O	$^{10}B + ^{10}C$	.135	11 $\pm$ 3	9	
		.60	12 $\pm$ 5	9	
		19	15 $\pm$ 8	9	
	$^{11}B + ^{11}C$	.135	25 $\pm$ 8	9	
		.60	25 $\pm$ 12	9	
		19	< 45	9	
Mg	^{7}Be	.60	6.5 $\pm$ 1.5	27	
		1.0	8.5 $\pm$ 1.0	38	
		2.0	10.2 $\pm$ 1.0	38	
		3.0	10.1 $\pm$ 0.9	38	
		23	9.9 $\pm$ 0.9	38	
	^{10}Be	.60	1.3 $\pm$ 0.4	27	
Si	^{7}Be	.60	5.3 $\pm$ 1.2	27	
		1.0	7.6 $\pm$ 0.8	38	
		2.0	10.8 $\pm$ 1.1	38	
		3.0	10.6 $\pm$ 0.9	38	
		23	10.7 $\pm$ 0.9	38	
	^{10}Be	.60	0.7 $\pm$ 0.3	27	
Fe	^{7}Be	1.0	4.0 $\pm$ 0.5	38	
		2.0	7.9 $\pm$ 0.8	38	
		3.0	9.4 $\pm$ 0.8	38	
		21	11.4 $\pm$ 1.2	4	
		23	11.7 $\pm$ 1.0	38	
	^{9}Be	21	8.1 $\pm$ 1.2	4	
	^{10}Be	.60	.44 $\pm$ .09	4	
		21	4.6 $\pm$ 0.8	4	
	^{45}Sc	.60	28.4 $\pm$ 1.8	4	
		21	18.0 $\pm$ 1.9	4	

Target	Product	Energy (GeV)	σ * (mb)	Refs	Notes
Fe	^{49}V	.60	37.5 ± 3.3	4	
		21	18.6 ± 3.2	4	
	^{50}V	.60	17.8 ± 1.3	4	
		21	10.0 ± 1.6	4	
	^{51}V	.60	6.8 ± 1.0	4	
		21	2.9 ± 0.6	4	
	^{50}Cr	.60	26.8 ± 3.5	4	
		21	15.1 ± 2.4	4	
	^{51}Cr	.60	43.3 ± 2.3	51	b
		21	25.1 ± 3.2	4	
	^{52}Cr	.60	69 ± 15	4	
		21	45.7 ± 11.6	4	
	^{53}Cr	.60	12 ± 2	4	
		21	8.5 ± 1.7	4	
	^{54}Cr	.60	3.5 ± 1.0	4	
		21	2.4 ± 1.0	4	
	^{53}Mn	.60	45 ± 4	4	
	^{54}Mn	.60	33.4 ± 1.6	51	b
		21	29.2 ± 2.7	4	
	^{55}Mn	.60	29 ± 7	4	
Ni	^{6}Li	3.0	17.3 ± 3.8	23	f
	^{7}Li	3.0	17.3 ± 3.8	23	
	^{7}Be	1.0	4.9 ± 0.6	38	
		2.0	9.3 ± 0.9	38	
		3.0	11.3 ± 1.0	38	
		23	14.0 ± 1.2	38	
	^{9}Be	3.0	4.3 ± 1.0	23	f

Target	Product	Energy (GeV)	σ * (mb)	Refs	Notes
Ni	^{10}Be	3.0	2.2 $\pm$ 0.5	23	
	^{10}B	3.0	5.6 $\pm$ 1.2	23	f
	^{11}B	3.0	8.6 $\pm$ 1.9	23	f

* Except for ^{7}Li or where specifically noted, all cross sections are cumulative (i.e. ^{6}Li + ^{6}He, ^{11}C + ^{11}B, etc..)

(a) Corrected for revised ^{10}Be half-life given in Ref. 50

(b) Adopted normalizing cross section

(c) Assumed normalizing cross section (see Ref. 15)

(d) Error adjusted to include present uncertainty in mass spectrometer discrimination factor for Be.

(e) Adjusted for improved estimate of σ_N (^{7}Be), based on Ref. 47

(f) These are " independent " (not cumulative) cross sections.

APPENDIX   TABLE   II

Some alpha induced cross sections relevant to cosmic ray spallation

Target	Product	Energy[**] (GeV/n)	σ[*] (mb)		Refs	Notes
^{12}C	^{6}Li	0.22	33	$\pm$ 5	11	
	^{7}Li	0.22	36	$\pm$ 6	11	
	^{7}Be	0.23	20.0 $\pm$ 1.0		48	
		0.70	16.3 $\pm$ 1.5		14	
		1.15	16.3		36	a
	^{9}Be	0.22	10.6 $\pm$ 1.7		49	
		0.70	10.0 $\pm$ 1.4		14	
		1.15	10.0		36	a
	^{10}Be	0.22	6.5 $\pm$ 1.4		49	
		0.70	5.4 $\pm$ 0.9		14	
		1.15	5.4		36	a
^{16}O	^{7}Be	0.70	16.3 $\pm$ 2.2		14	
		1.15	16.3		36	a, b
Mg	^{7}Be	0.70	16.7		36	b
		1.15	17.7		36	a, b
Si	^{7}Be	0.70	16.9		36	b
		1.15	19.1		36	a, b
Fe	^{7}Be	0.70	15.5		36	b
		1.15	20.3		36	a, b
	^{51}Cr	0.70	40.4		36	b
		1.15	38.8		36	a, b
	^{54}Mn	0.70	45.0		36	b
		1.15	45.3		36	a, b

Target	Product	Energy[**] (GeV/n)	σ[*] (mb)	Refs	Notes
Ni	^{7}Be	0.70	18.1	36	b
		1.15	26.4	36	a, b
	^{51}Cr	0.70	33.5	36	b
		1.15	35.9	36	a, b
	^{54}Mn	0.70	15.8	36	b
		1.15	18.3	36	a, b

[*] Except for ^{7}Li or where specifically noted, all cross sections are cumulative (i.e. ^{6}Li + ^{6}He, ^{11}C + ^{11}B, etc..)

[**] This energy corresponds approximately to that (in GeV/n) of cosmic ray progenitor interacting with interstellar helium.

(a) Cross sections calculated assuming σ_c (^{7}Be) at 1.15 GeV/n is the same as at 0.7 GeV/n (see Ref. 36)

(b) Errors not quoted because results are still preliminary. Estimated uncertainty $\sim$ 15-20 %.

THE LIGHT ELEMENT FORMATION: A SIGNATURE OF HIGH ENERGY NUCLEAR ASTROPHYSICS

J. Audouze[*]

McDonnell Center for Space Sciences, Washington
University, Saint Louis, Missouri 63130

and

M. Meneguzzi[+] and H. Reeves

Division de la Physique SES
CEN, Saclay, France

SUMMARY. Light elements D, ^{6}Li, ^{9}Be, ^{10}B and ^{11}B (and possibly also ^{7}Li) are not produced by the general nucleosynthetic processes occurring in stars. They appear to be synthetized by high energy processes occurring either during the interaction of galactic cosmic rays with the interstellar medium or in supernovae envelopes. These formation processes are discussed. It is emphasized that the most coherent scenario regarding the formation of the light elements is obtained by taking also into account the nuclear processes which may have occurred during hot phases of the early Universe (Big Bang). Implications on chemical evolution of galaxies and on cosmology are briefly recalled.

I. INTRODUCTION

Since about two decades ago (Fowler et al., 1955; Burbidge et al., 1957), light elements such as D, Li, Be and B have been known to have special nucleosynthetic properties: they are not formed but destroyed by the thermonuclear reactions occurring

---

[*]Current address: Observatoire de Meudon, Meudon, France and Laboratoire René Bernas, Orsay, France.
[+]Also at DAF, Observatoire de Meudon, Meudon, France.

Shen/Merker (eds.), Spallation Nuclear Reactions and Their Applications, 113–137. All Rights Reserved.

inside the stars. One exception, however, must be made regarding
^{7}Li: Cameron and Fowler (1971) [see subsequent works by Ulrich
and Scalo (1972) and by Sackmann et al. (1974)] have proposed
that this particular element could be synthetized in red giant
stars during the mixing of hydrogen-rich and helium-rich zones
by the reaction ^{3}He$(\alpha,\gamma)^7$Be, provided that ^{7}Be is efficiently
transported by convective motions at the surface of these stars[1]
This process would explain the rather large ^{7}Li abundance,
^{7}Li/H $\sim$10^{-7}, observed in some red giants (Boesgaard, 1970).
Furthermore, some recent calculations of explosive nucleosynthesis
in hydrogen and helium burning zones have been performed which
seem to show that ^{7}Li, and to a less extent ^{11}B, could be produced by such processes (Arnould and Nogaard, 1975; Nogaard and
Arnould, 1975; Toussaint, 1975).

It is now believed (at least for ^{6}Li, ^{9}Be and the boron
isotopes) that these elements come from the spallation reactions
that carbon, nitrogen and oxygen might suffer in special astrophysical events where these high-energy reactions take place.
Up to 1970, it was generally assumed that these spallation reactions occur at the surface of young and very active objects,
the T Tauri stars, where the upper lithium abundances were observed. This hypothesis has been defended, for instance, by
Bernas et al. (1967) and presented by Mitler (1967) in the proceedings of the previous Philadelphia conference.

In 1970, Ryter et al. realized that these spallation reactions, occurring at the surface of young stars and extracting
their energy only from gravitational sources, demand amounts of
energy larger than these stars can effectively release. This
drawback of the "autogenetic"[2] hypothesis induced Reeves et al.
(1970) to follow an unpublished suggestion of B. Peters and to
propose that the bulk of the light elements are formed by the
interaction of the galactic cosmic rays with the interstellar
medium, leading more naturally and with less energy expenditure
to the spallation reactions producing these elements. Then
Meneguzzi et al. (1971) (hereafter quoted as MAR) performed more
detailed calculations in the framework of such a "galactogenetic"[2]

---

[1] See however Iben (1973).

[2] The term, autogenetic applies to a process which occurs in a
specific star. It has been opposed by Reeves (1971) to the term
galactogenetic, which is applied to a process occurring on a universal or galactic scale such as that induced by cosmic rays in
the interstellar medium.

model and showed that ^{6}Li, ^{9}Be, ^{10}B and ^{11}B are naturally formed
by this process. Similar calculations and conclusions have been
also reached by Mitler (1972).

Problems remained with the formation of D and ^{7}Li, which are
not formed in sufficient quantities by the observed galactic cosmic rays (hereafter called GCR). Many different solutions, more
or less successful, have been proposed which will be reviewed
here.

An outline of this review of the question regarding the
nucleosynthesis of these elements and related fields such as the
chemical evolution of galaxies and some cosmological consequences
follows. In Section II we reanalyze briefly the relevant recent
observations regarding especially deuterium and boron which are
still now subject to discussion. For the remaining body of observational data the reader is referred to Reeves et al. (1973),
Schramm and Wagoner (1974) and especially Reeves (1974a). In
Section III we summarize briefly the nucleosynthetic properties
of the GCR regarding the light elements, following mainly the
analysis presented by MAR. Then several solutions regarding the
production of suprathermal (up to a few tens MeV) particles are
analyzed in Section IV: low energy GCR as proposed by MAR,
Audouze and Truran (1973) and more recently by Meneguzzi and
Reeves (1975), Bodansky et al. (1975) and Canal et al. (1975);
the spallation reactions induced in supernovae envelopes (or in
supermassive stars, see e.g. Hoyle and Fowler, 1973) proposed
first by Colgate (1973, 1974, 1975) and then discussed or re-
analyzed, for instance, by Reeves (1974b), Epstein et al. (1974),
Weaver and Chapline (1974) and Bodansky (1975). It must be emphasized also that Canal (1974) and Canal et al. (1975) have
rediscussed the point that some reactions such as the $\alpha + \alpha$
reaction might occur at the surface of young stars and found
some arguments which decrease the energetic requirements over
those estimated by Ryter et al. (1970). In the last section
(Section V) we present the problem of the formation of these
elements in a framework somewhat more general than the simple
consideration of spallation reactions by examining the possible
consequences of some "Big Bang" nucleosynthesis regarding the
formation of these elements and by presenting our preferred
scenario and making a few comments regarding the relevance of the
light element nucleosynthesis to some related questions such as
the evolution of the galaxies and cosmology.

II.   ADDENDA ON THE DEUTERIUM AND BORON ABUNDANCE

As said before, the reader is referred to Reeves (1974a)
for a rather complete survey of the observations of the light
element abundances up to 1973. Here, we will only reproduce the

Reeves (1974a) tables (Table 1; see pp. 131-135). We believe however
it is useful to rediscuss the latest status of the situation regarding D and B which play a rather important role in this presentation.

1. Deuterium

It seems now that the D/H ratio is approximately uniform in
the solar system, which is some material that has been isolated
4.6×10^9 years ago. $\frac{D}{H} \sim 2 \times 10^{-5}$, according to measurements in
Jupiter's atmosphere (Beer and Taylor, 1973), the analysis of He
isotopic ratios in the solar wind (Geiss and Reeves, 1972) and
the measurement of $\frac{D}{H}$ in the nearby interstellar medium. Thus,
Rogerson and York (1973) have found $\frac{D}{H} = 1.4 \times 10^{-5}$ by analyzing
Copernicus observations of the Lyman features in the line of
sight to β Cen. Similarly, York and Rogerson (1975) measuring
along the line of sight to μ Col, γ^2 Vel, α Cru AB and α Vir AB
propose $\frac{D}{H} = (1.8 \pm 0.4) \times 10^{-5}$, a value they believe represents
the present composition of the region within 200 pc of the sun.
This result has been confirmed by other Copernicus measurements
in the lines of sight of ζ Pup and γ Cas but analyzed with different techniques by Vidal-Madjar et al. (1975)

2. Boron

A recent paper by Cameron et al. (1973) led to a rather
complicated puzzle concerning the abundance of this element.
These authors have claimed that the interstellar abundance of
boron must be $\frac{B}{H} \sim 10^{-8}$ based upon carbonaceous chondrites. In
fact, Morton et al. (1974) have set a rather stringent limit on
this ratio ($\frac{B}{H} < 7 \times 10^{-11}$), indeed more stringent than that obtained
by Audouze et al. (1973) analyzing the first Copernicus data and
confirmed very recently by Steigman (1975)[1]

The situation has become rather clear now, especially with
the new measurement of the B abundance in the atmosphere of a
rather young normal A star α Lyrae, performed by Boesgaard et al.
(1974) and giving $\frac{B}{H} \sim 8 \times 10^{-11}$, also consistent with a recent
upper limit put on the solar abundance of $\frac{B}{H} < 10^{-10}$ by Hall and
Engvold (1975). We would then be inclined to adopt a ratio of
$\frac{B}{H} \sim 10^{-10}$, a value that Cameron (private communication) is
willing to accept now.

---

[1] It might be argued that boron has been locked in grains as other
heavy elements, as proposed, for instance, by Field (1974). But
any boron molecule has a condensation temperature similar to that
of S which seems to be rather undepleted after Morton (1974) and
York (1975).

To end this section, we have listed in Table 2 (see p. 136) what we would consider as the "best" cosmic abundances for the light elements.

III.   THE "CANONICAL" GALACTIC COSMIC RAY MODEL

The "canonical" GCR model designates here the evaluation of the production of light elements produced by the present observed flux of GCR integrated throughout the life of the galaxy: the main observational feature which has prompted Reeves et al. (1970) to propose this mechanism is that the L/M ratio L = LiBeB; M = CNO nuclei) is about 0.23 in the GCR while it is only a few 10^{-6} in the "standard abundances". For the description of this model, we summarize the procedure and the results presented by MAR.

In this work, the study of the interaction of the GCR with the interstellar gas has been made by using a rather simple but presumably reasonable approach. The galaxy is supposed to be a "leaky" box in which the sources of GCR (we would say the supernovae) are distributed homogeneously. These sources emit continuously a time-independent flux of high-energy particles. The expression "leaky box" means that GCR particles have some probability of leaving the galaxy when they reach its border. The possible events that a GCR particle can experience are schematized in Fig. 1. Besides the escape, the particles can induce spallation reactions with the nuclei of the interstellar

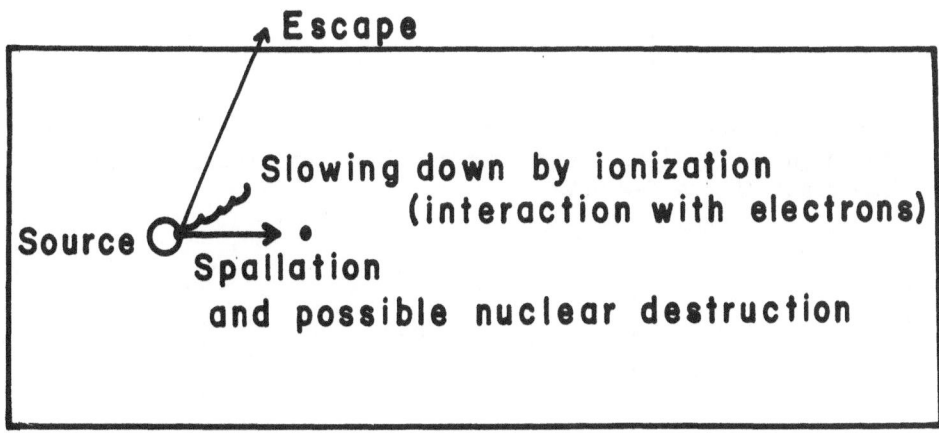

Fig. 1. Sketch of the fate of the GCR particles interacting with the interstellar medium.

atoms and are slowed down by inelastic collision with the electrons of these atoms. If one assumes that the GCR flux has reached a steady state between all these effects, an equation describing this steady state can be simply written:

$$(1) \quad \frac{\partial N_i(E)}{\partial t} = 0 = -\frac{N_i(E)}{\tau_e} - \frac{N_i(E)}{\gamma \tau_i} + Q_i(E) + \frac{\partial}{\partial E}\left(b_i(E)N_i(E)\right)$$

$$-\left[\sigma_{\alpha i}(E)n_{He} + \sigma_{pi}(E)\, n_H\right] v_i N_i(E)$$

$$+\sum_j \int_0^\infty N_j(E')\left[n_{He}\sigma_{\alpha ji}(E,E') + n_H\sigma_{pji}(E.E')\right]v_j dE'$$

In this equation $N_i(E)$ represents the number of GCR particles of type i of energy E per unit of volume and time. On the right-hand side of equation (1) the first term represents the probability of escape of the particles i out of the galaxy (τ_e is the lifetime against escape); the second term represents the probability that a radionuclide i suffers a beta-decay; Q_i is the source term of GCR particles i, $b_i(E) = -\left(\frac{\partial E}{\partial t}\right)_i$ expresses the energy loss per nucleon per sec by ionization (or more generally by inelastic collisions with electrons), $\sigma_{\alpha i}$ and σ_{pi} are the total destruction cross sections while $\sigma_{\alpha ij}$ and σ_{pij} are the spallation cross sections for the production of i by bombardment of j by alpha-particles or protons. n_H and n_{He} are the hydrogen and helium densities.

This rather simple model corresponds to a distribution of escape path length X expressed in g cm^{-2} which is exponential

$$(2) \quad P(X) = \frac{1}{\Lambda}\exp\left(\frac{-X}{\Lambda}\right) ,$$

where the escape mean free path Λ can be estimated simply from the high-energy fluxes when the deceleration processes are not important (E > a few GeV)

$$(3) \quad \frac{1}{\Lambda} = \sum_i \frac{\phi_i}{\phi_j}\left[\sigma_{pij} + \sigma_{\alpha ij}\frac{n_{He}}{n_H}\right] - \frac{\sigma_{pj} + \sigma_{\alpha j}\frac{n_{He}}{n_H}}{M_p + \frac{n_{He}}{n_H}M_\alpha} .$$

In this expression the ϕ are the GCR fluxes ($\phi = Nv$), M_p and M_α the masses of the hydrogen and helium atoms. Λ has been found to be equal to ~ 6 g cm^{-2}.

The light elements are then produced by spallation induced
(a) either by the rapid GCR protons and alpha particles bombarding the interstellar CNO atoms, or (b) by the reverse process.
In the latter case the light element nuclei have a velocity
comparable to that of their GCR progenitors (CNO) and have to
slow down before they can be considered as interstellar atoms
themselves. Contributions (a) and (b) can then be summed and
the number of light element nuclei i produced per unit of time is:

$$(4) \frac{\partial n_i}{\partial \tau} = \sum_j \int_0^\infty \left[\sigma_{pji}(E')\phi_p(E') + \sigma_{\alpha ji}(E')\phi_\alpha(E) \right] n_j dE' + N_i(E)b_i(E).$$

MAR have made the calculations by using the observed GCR
fluxes $\phi_p(10 \text{ GeV}) = 2.5 \times 10^{-2}$ protons $\text{cm}^{-2}\text{sec}^{-1}\text{GeV}^{-1}$ and an
injection spectrum $q \propto W^{-2.6}$ where W is the total energy per
nucleon. The contributions of processes (a) and (b) represent
respectively ~ 0.7 and ~ 0.3 of the total formation rate. The
results of their computations are given in Table 3 (see p. 137).

For these calculations it is recalled that MAR have used a
total energy injection spectrum without assuming in this case
the presence of huge fluxes of low-energy particles up to a few
tens of MeV, which remain near the sources since they are rapidly
slowed down or are prevented to enter the solar cavity due to
solar modulation effects. The assumption of the presence of such
fluxes is examined below.

From Table 3, integrating the present GCR flux over 10^{10}
years, which is a rough estimate of the assumed age of the
Galaxy, one can see that GCR are likely to produce the observed
^{6}Li, ^{9}Be, ^{10}B, and ^{11}B, while Li is underabundant by about a
factor of 10. Calculations on D and ^{3}He have not been reproduced.
In this case the underabundance reaches factors of about ~ 100.

To conclude this section, with the present observed GCR
fluxes integrated over the age of the Galaxy one easily explains
the formation of ^{6}Li, ^{9}Be, ^{10}B and ^{11}B but not that of D, ^{3}He
and ^{7}Li.

IV. SPALLATION PROCESSES INDUCED BY LOW ENERGY PARTICLES

To try to solve the above difficulties encountered when only
the high energy GCR flux is considered, effects of low energy
particles (up to a few tens of MeV) have been considered by
various authors in different astrophysical contexts. The use of
such low energy particles is motivated by several reasons: (i)
as stated before, it is quite likely that acceleration processes
occurring in GCR sources produce a very copious amount of such

suprathermal particles but these particles cannot go very far from their sources due to the strong energy loss they suffer (the stopping power is inversely proportional to the energy), and they may experience solar modulation if by any chance they reach regions close to the solar system; (ii) the threshold effects are such that the productions of ^{7}Li and ^{11}B are favored compared with those of the other isotopes due to the lower thresholds of the spallation reactions producing them.

Two papers, Audouze and Truran (1973) and Bodansky et al. (1975) (who included reactions induced on ^{13}C), have based their analysis mainly on the use of the Q-value effects and have shown that large fluxes of $\sim$10-15 MeV particles are able to produce the ^{7}Li (and to some extent ^{11}B) needed to explain the observed ^{7}Li/^{6}Li ^{11}B/^{10}B or Li/B or Be. These studies support the idea that huge low-energy fluxes (of any type: GCR, flare particles or particles produced during the outward propagation of the shock wave in the atmosphere of a supernovae) can help solve the ^{7}Li/^{6}Li ratio and ^{11}B/^{10}B ratio problem (^{7}Li/^{6}Li = 12.5 and ^{11}B/^{10}B = 4 in the solar system while MAR calculations based upon the use of fluxes in power law of the total energy lead to ^{7}Li/^{6}Li $\sim$ 2 and ^{11}B/^{10}B = 2.8). It must be said, however, that such investigations suffer the drawback of not following very properly what happens to a suprathermal particle during their deceleration and can really only apply to spallation mechanisms occurring during acceleration processes (confinement phases) themselves. This difficulty is also encountered by Hainebach et al. (1975a,b) who pursued an earlier work of Audouze (1970) and try to explain, together with the light elements, the formation of rare odd-odd nuclei such as ^{50}V, ^{138}La and ^{180}Ta.

In somewhat more elaborate calculations MAR and more recently Meneguzzi and Reeves (1975) have simply assumed the presence of low energy GCR fluxes compatible with the observed heating rate of the interstellar medium. Since the low energy particles are easily slowed down, they constitute one of the main ionization agents of the interstellar medium. These two works used the same formalism as expressed in the previous section: MAR assumed a flux between 5 to 30 MeV/nuc.$\propto 10^3 W^{-2.6}$ (they designated it as the low-energy GCR "carrot"); this flux is in agreement with the heating rate of the interstellar gas and leads to Li/H $\sim 10^{-9}$ and B/H $\sim$ few times 10^{-10}, but to ^{7}Li/^{6}Li $\sim$2 which is not the actually measured ^{7}Li/^{6}Li ratio ($\gtrsim$ 10 in any astrophysical site). For this reason, Meneguzzi and Reeves (1975) have played with various parameters such as the spectral shape of the carrot (assumed to be $E^{-\gamma}$ with γ varying from 3 to 7), the value of the escape length and also the composition of the interstellar gas. They found that if the escape length is not large ($\Lambda < 3$ g cm^{-2}) and if the low GCR spectrum is rather soft ($\gamma \sim 3$) low energy fluxes could explain large values of ^{7}Li/^{6}Li and Li/Be. The conditions are, of course, less

stringent if there is an enrichment of the CNO elements in the
source. They therefore reach the conclusion that neither the
energetic arguments, nor the present heating rates, nor the gamma-
ray production in the sources producing such huge low-energy GCR
fluxes can really rule out the presence of these fluxes and the
possibility of explaining large Li/Be and ^{7}Li/^{6}Li ratios this way.
They expect that improvements in gamma-ray detection or a better
knowledge of the heating of the interstellar medium[1] will permit
a definitive conclusion on the occurrence of such huge low-
energy GCR.

Similar calculations which take into account [as already
calculated by Audouze (1970)] possible destruction of the light
elements have been performed by Canal et al. (1975), who reach
quite similar but slightly more optimistic conclusions regarding
the possibility of low-energy GCR fluxes.

It must be remembered that all these models are unable to
produce sufficient amounts of deuterium. Since about two years
ago, Colgate (1973, 1974, 1975) has elaborated and advocated a
very clever model but which is not free from difficulties. The
basic ideas under his model are the following: when a shock wave
propagates in a supernova envelope, the amount of thermal energy
which is transported through its passage is shared by fewer and
fewer particles since the density decreases when the shock wave
propagates towards the surface of this object. For densities
such as $10^{-8} < \rho < 10^{-5} g$ cm^{-3} the Colgate model predicts that a
fair fraction of the particles can reach energies higher than
$\gtrsim 30$ MeV/n, leading to the complete break-up of ^{4}He into protons
and neutrons. The density conditions are such that instead of
decaying the neutrons recombine with protons to give D during the
cooling process. The cooling which occurs after the passage of
the shock wave is due to the transfer of the energy from the ions
to the electrons by ionization processes, then to photons by
inverse Compton and bremsstrahlung mechanism. As we shall discuss
later, the occurrence of this cooling process is one of the
critical parameters of the Colgate model. According to this
author, the cooling is rapid but not enough to prevent a fair
fraction of the ions to reach the energy needed to spallate completely ^{4}He. If such a mechanism is actually working, one single
source, i.e., supernova, can explain via the low-energy and the
high-energy particles all the light elements at once. A quite
similar physical model but working in the super-massive stars

---

[1] The situation in this respect now is rather confused. There are
calculations such as those of Meszaros (1974) which would conclude
that the ionization rate of the interstellar medium is rather low,
thus ruling out such low-energy fluxes. But these calculations are
based upon stationary models which may not apply to the real
situation [see e.g. Gerola et al. (1974) and Field (1975)] .

has independently been proposed by Hoyle and Fowler (1973).

To come back to the Colgate mechanism, one must say that many criticisms coming from different arguments have been presented. They can be listed under three headings: (i) the cooling time-scale, (ii) the energetics requirements, and (iii) the actual products of the nucleosynthetic processes.

(i) The cooling time-scale - Weaver and Chapline (1974) claimed that the treatment of the cooling process by Colgate has been incomplete. Besides the radiative equilibrium occurring after the passage of the shock wave, other cooling effects play an important role in fixing the temperature decrease. The photons diffuse on the electrons and then radiate an enormous amount of energy. Some energy is also released by electron-electron bremsstrahlung radiative Compton diffusion, inverse Compton scattering and pair creations. When all these effects are properly taken into account, they found that the energy W appearing in (5) (see later) is less than 100 keV making the Colgate model unable to produce significant fractions of deuterium.

(ii) The energetic requirements - Reeves (1974b) has presented some arguments tending to rule out the Colgate model on the following grounds: according to the Colgate model the amount of deuterium which is produced by the passage of such shock waves is:

$$(5) \qquad\qquad \frac{n(D)}{n(H)} \sim 10^{-4} \, W^2 .$$

In this expression W represents the average energy which is deposited by the shock wave. In order to make the Colgate model work an average energy W of $\sim$1 MeV/n is needed; but the Kirshner et al. (1973) measurements of the velocity of the material ejected from a supernova remnant leads to W$\leq$0.2 MeV/n (Colgate (1974) indeed disagrees with this use of the present supernova observations). Moreover, if one can take 10^{51} ergs as the maximum energy released by a supernova and a mass for the envelope which is expanding of $\sim$5 M$_\odot$, one gets D/H$\sim$4x10^{-6} in any given object. Since the mass fraction which has been involved in exploding objects is certainly less than 10%, (D/H)$_{SN}$ <4x10^{-7}, i.e. $\sim$2% of what is really needed to explain the observed D/H abundance.

(iii) The nucleosynthetic processes - Even if a value of W as large as 30 MeV is correct in his papers, Colgate has not tried to make a systematic analysis of all the possible nucleosynthetic mechanisms which can accompany the deuterium production. It is indeed obvious (and that was the hope of Colgate too) that the other light elements might be produced by the spallation reactions involving the C, N and O nuclei. Such an analysis has been undertaken by Epstein et al. (1974). As a result of their analysis they have shown that the required production of D would be accompanied by an overproduction of ^{7}Li, ^{9}Be and ^{11}B by factors ranging from 5 to 30. A similar but simpler analysis is currently being done by Bodansky (1975), who has noticed that the treatment of the interaction of ^{7}Li with neutrons might have not been made satisfactorily by Epstein et al. (1974)[1] He argues that neutrons do not lead to the destruction of the mass 7. He concludes his analysis by saying that if, contrary to the conclusions of Weaver and Chapline (1974), the temperature of the region experiencing the shock passage is very large, the ratio Li/D might not be larger than the observed one. However, one other drawback of his simple analysis lies in the fact that he has considered material deprived of CNO. The arguments concerning ^{9}Be and ^{11}B put forward by Epstein et al. (1974) still constitute a major difficulty for the Colgate mechanism.

It must be noted that the consideration of the Colgate model is very interesting and has been so far very fruitful in the understanding of the outburst of the shock waves in the exterior of exploding objects: even if D or light elements are not produced this way, the Colgate mechanism can be very well used to study still poorly-known nucleosynthetic processes such as the study of the p process undertaken by Truran and Cameron (1975); see e.g. Truran (1973) for a summary of their work. However, in view of such difficulties it does appear now that the contribution of the shock wave nucleosynthesis to cosmic deuterium and other light elements such as Li Be B is negligibly small.

Before closing this section devoted to spallation reactions induced by low-energy ($\sim$a few tens of MeV) particles, one must note that a few recent works try to give a new life to "autogenetic" processes. In particular, Canal (1974) has argued that the reservation made by Ryter et al. (1970) on the "autogenetic" production of ^{7}Li in T Tauri stars does not really apply to the $\alpha + \alpha$ reaction rate since He is $\sim$ 100 times more abundant than CNO and the $\alpha + \alpha$ cross section is rather large at low energies. He has then suggested that ^{7}Li could indeed be produced by flare stars. He has also based his arguments on the fact that lithium has been found to have a larger abundance $\sim 10^{-9}$ in T Tauri and younger stars than in the interstellar medium (if one takes, for instance, the Traub and Carleton (1973) value of Li/H $\sim 3 \times 10^{-10}$ at its face value). Similar conclusions based on an analysis treating the light element (LiBeB) problem in more general terms have been proposed in a subsequent paper by Canal et al. (1975). But, although their point regarding the ^{4}He + ^{4}He reaction rate

[1]See however Epstein et al. (1975).

is correct, their model remains still doubtful: (i) the high-
energy GCR produce indeed the correct amount of ^{6}Li, ^{9}Be and B
and their model can apply only to ^{7}Li; (ii) the difference between
the interstellar medium ^{7}Li abundance and that observed in young
stars is not significant if we give some weight to the measurement
of Vanden Bout and Grupsmith (1973), Li/H $\sim$ 6 x10^{-10}, and one may
argue that the difference between the interstellar medium and the
young stars can be explained by the fact that part of the Li atoms
are going into grains; (iii) furthermore, an "autogenetic" model
might have trouble in explaining as large a Li/H upper abundance
as $\sim$10^{-9}, as shown by Zappala (1972). For these reasons, although
this argument of a possibility of "autogenetic" production of light
elements cannot be ruled out entirely, we take the view that
galactogenetic processes are more important as far as the production of these elements are concerned.

    Along the same line of arguments, Ramadurai and Wickramasinghe
(1975) have speculated that the high boron abundances observed in
carbonaceous chondrites (B/H $\sim$ a few times 10^{-9}) might be explained
in terms of rather profuse irradiations by energetic particles
(fluxes of $\sim$1.5 x. 10^{20} particles cm^{-2} of energy > 20 MeV are
needed to account for such high boron abundances). This speculation seems very dubious since one would expect some large differences between the ordinary chondrites and the carbonaceous
chondrites not only as regards Be (as already recognized by the
authors themselves) but also as regards many other trace elements
such as the light isotopes of Xe (^{124}Xe, ^{126}Xe which are very rare).

    To close the discussion about the light element production by
low-energy particles, it appears that it is almost impossible to
produce D by such fluxes. On the other hand, it solves rather
efficiently the problem of ^{7}Li following, for instance, the
Meneguzzi and Reeves (1975) work.

       V. ORIGIN OF LIGHT ELEMENTS, BIG BANG,
            COSMOLOGY, CHEMICAL EVOLUTION

    If one wants also to explain the formation of D together
with that of LiBeB, it seems rather likely that some other source
needs to be investigated. It has been known since 1966 after some
calculations made first by Peebles, substantiated by Wagoner et al.
(1967) and reexamined more recently by Wagoner (1973), that D and
also ^{3}He and ^{7}Li can be formed easily in the nucleosynthetic
processes occurring in the early Universe. As can be seen, for
instance, from Wagoner (1973), who studied mainly the so-called
"canonical" Big Bang (we will define this term in the following
paragraph), all the D which is seen now can be produced if the
present density of the Universe is less than 5 x 10^{-31}g cm^{-3}

(with T_{now} = 2.7°K). In this way all the ^{3}He is also explained while only $\sim$10% of the observed ^{7}Li is produced.

To define what we call the "canonical Big Bang", one makes a series of assumptions about the cosmological model in which the early phases of the Universe are supposed to be described. We adopt here the presentation of Wagoner (1974) and Schramm and Wagoner (1974). Two assumptions can be considered as fundamental: (i) the Universe is homogeneous and isotropic, and (ii) the principles of special relativity are locally valid in all freely falling reference frames. These two assumptions seem to be in rather good agreement with all the present body of observational data. In contrast, the remaining hypotheses may be more easily violated by the real universe. In the "canonical" Big Bang one assumes also: (iii) the temperature was rather high at the be-ginning (T >10^{11}°K) to allow statistical equilibrium among all the particles present, (iv) only known particles were present during the nucleosynthesis; in particular, there were no quarks, super-baryons or other strange beasts; one assumes also that the primordial magnetic fields were negligible, (v) all the particles were non-degenerate (this applies to the neutrinos), (vi) the baryonic number was non-zero (not considering here symmetric models as those produced by Omnès)[1]; (vii) finally, one assumes that the general relativity theory applies and also that the expansion rate v is given by $\frac{1}{v}\frac{dv}{dt}$ = $\sqrt{24\pi G\rho}$, i.e., governed by the free fall time of the matter v is the velocity of a co-moving volume element and ρ is the total mass energy density.

The "canonical" Big Bang is then the model where all these assumptions apply and then in this case one gets the abundances given above, the conditions for getting enough D being that the present density of the Universe is much lower than the critical density.

Now this simple picture can of course be altered very easily in many ways:

1. The first alteration is to assume as Wagoner (1974) has done, for instance, that the expansion rate may have varied from the expression given above and could be described, for instance, by $\frac{1}{v}\frac{dv}{dt}$ = $\xi\sqrt{24\pi G\rho}$, with an adjustable parameter ξ. In particular, it can be shown that, if ξ = 2, in almost all the cases of inter-est D and ^{7}Li are produced together, ξ = 2 meaning that the time scale is smaller than the free fall time or that the expansion of the Universe occurs very rapidly.

---

[1]The attempt of building a symmetric model by Omnes and his col-laborators have run into severe difficulties in accounting for the cosmic helium abundance and for this reason the model can essen-tially be ruled out (Ramani Thesis).

2. The above (Wagoner, 1973) calculations have been made assuming a leptonic number equal to zero. Reeves (1972) has discussed the possibility of some models where this condition is relaxed. In some cases ^{7}Li can be made more easily together with D and ^{3}He (taking, for instance,

$$L_e = \frac{n_e^+ - n_e^- + n_\nu^- - n_\nu^-}{n_\gamma} = -0.18 \text{ and } \rho \sim 10^{-30} \text{ gm/cm}^3)$$

3. Epstein and Petrosian (1975) have examined the possibility of some density fluctuations occurring during the Big Bang, which according to these authors relax somewhat the constraint on the present density of the Universe for D production. In particular they show that, in their model, production of both ^{7}Li and D requires large amplitude fluctuations.

4. Goret and Beaudet (1975) and Beaudet and Yahil (1975) have recently investigated the effects of a large universal muon leptonic number on the Big Bang production of deuterium and helium. Yahil and Beaudet (1975) have shown that a Big Bang origin of deuterium does not rule out a closed universe contrary to the conclusion of Reeves et al. (1973) or Gott et al. (1974) , provided that the universe is filled with muon neutrinos (or any massless particles) with strong matter-antimatter asymmetry.

Although the problem of the origin of the light elements has remained after so many years and so many written and spoken comments (a small fraction of the light elements may come from an autogenetic origin; an even smaller fraction of deuterium may come from shock wave processes), it does seem highly plausible that the bulk of D, ^{3}He and ^{4}He come from the Big Bang and that the bulk of ^{6}Li, ^{9}Be, ^{10}B and ^{11}B come from the GCR bombardment of the interstellar medium.

^{7}Li is still highly problematic. Galactic cosmic rays, including a hypothetical low-energy component which is still unobserved and therefore of unknown flux, will account for a fraction of the ^{7}Li, although we find it difficult to say that they could go all the way to explain the meteoritic ratio of ^{7}Li/^{6}Li = 12.5. A definite answer could come from a better understanding of the heating and ionization mechanism of the interstellar medium and from detection of nuclear gamma rays in the vicinity of cosmic ray sources (supernovae remnants ?), as discussed in Meneguzzi and Reeves (1975).

Big Bang nucleosynthesis does contribute to ^{7}Li but again not nearly sufficiently, unless particular and rather ad hoc assumptions are made.

Finally, stellar wind ejection from red giants is another

possible contributor although we are unable now to give any quantitative estimate.

To end this review, it must be noted that the light elements play a very important role in building models of chemical evolution. This has been shown in particular by Reeves et al. (1973), Audouze and Tinsley (1974, 1976), Reeves (1974a) and Tinsley (1974). This is due to the specific properties of the light elements which have been pointed out at the beginning of this review. They are not produced like the other elements by the nucleosynthetic processes occurring in the interior of stars. On the contrary, when they go into a star they are easily destroyed. This is true in particular of D and ^{6}Li, which can be used as parameters to measure the amount of material which has been "astrated", i.e., which has been at least processed once into a star. The fact that D has not varied much during the last 4.6 x 10^9 years tells us that, if D comes from the Big Bang, either the astration has been small which is somewhat contradicted by some considerations on the evolution of the ^{12}C/^{13}C ratio; see Audouze et al. (1975) or some infall of intergalactic gas of primordial composition occurs in the galactic disk to prevent the strong depletion of D which might be due to astration processes.

In light of current evolution models, Ostriker and Tinsley (1975) have proposed a way to disentangle the mystery of the deuterium origin. They show that D must increase with the metal abundance Z if the Colgate model works, or decrease with Z if D is produced by the Big Bang. Although the idea is interesting, contrary to their hopes measurements of the deuterium abundance far from the solar neighborhood where the metal abundance can be different must be terribly difficult and are not feasible in the foreseeable future. Finally, light elements can be used to give some idea of the flux of galactic cosmic rays, then on the evolution of rate of supernova explosions, then indirectly on the initial mass distribution of stars. The differences in their destruction mode can also give some insight on the way that material is processed into stars. The reader is referred to the above references for some specific studies of these still intriguing questions.

The problem of the origin of the light elements still remains a very exciting puzzle. One can say that some progress has been made in this field in the last twenty years where this question has been studied. However, it has been obvious throughout this review that many uncertainties, many unknowns, still leave much obscurity on a question which has so many implications on nuclear astrophysics. That must be considered presumably as an advantage for people who have worked and are still working to try to understand the nucleosynthesis of such elusive elements.

ACKNOWLEDGMENTS

One of us (J.A.) would like to express his gratitude to Dr. Robert M. Walker and the other members of the McDonnell Center for the Space Sciences for their hospitality during the period when this communication was prepared and presented. He thanks also Dr. Martin Rees and the Institute of Astronomy at Cambridge (U.K.) for their hospitality during the period when the review was completed and written, and Ms. Hilda Ketterer for her help in the preparation of the manuscript. The survey of the literature was completed in August 1975.

REFERENCES

1. Arnould, M. and Nogaard,H., 1975, Astron. Astroph. In press.
2. Audouze, J., 1970, Astr. Ap. 8, 436.
3. Audouze, J., Lequeux, J. and Reeves, H., 1973, Astr. Ap. 28 85.
4. Audouze, J., Lequeux, J., Reeves, H. and Vigroux, L., 1976, submitted to Ap. J. Letters.
5. Audouze, J., Lequeux, J. and Vigroux, L. 1975, Astr. Ap. 43,71.
6. Audouze, J. and Tinsley, B. M., 1974, Ap. J. 192, 487.
7. Audouze, J. and Tinsley, B. M., 1976, Ann. Rev. Astr. Ap. 14.
8. Audouze, J. and Truran, J. W., 1973, Ap. J. 182, 839.
9. Beer, R. and Taylor, F. W., 1973, Ap. J. (letters) 182, L131.
10. Bernas, R., Gradsztajn, E., Reeves, H. and Schatzman, E. 1967, Ann. Phys. (N.Y.) 44, 426.
11. Bodansky, D., 1975, private communication.
12. Bodansky, D., Jacobs, W. W. and Oberg, D. L., 1975, Ap. J. 202, 222.
13. Boesgaard, A. M., 1970, Ap. J. 161, 1003.
14. Boesgaard, A. M., Praderie, F., Leckrone, D. S., Faraggiana, R. and Hack, M., 1974, Ap. J. (letters) 194, L143.
15. Burbidge, E. M., Burbidge, G. R., Fowler, W. A. and Hoyle, F. 1957, Rev. Mod. Phys. 29, 547.
16. Cameron, A. G. W., Colgate, S. A. and Grossmann, L., 1973, Nature 243, 204.
17. Cameron, A. G. W. and Fowler, W. A., 1971, Ap. J. 167, 111.
18. Canal, R., 1974, Ap. J. 189, 531.
19. Canal, R., Isern, J. and Sanahuja, B., 1975, Ap. J., in press.
20. Colgate, S. A., 1973, Ap. J. (Letters) 181, L53.
21. Colgate, S. A., 1974, Ap. J. 187, 321.
22. Colgate, S. A., 1975, Ap. J. 195, 493.
23. Epstein, R. I., Arnett, W. D. and Schramm, D. N., 1974 Ap. J. (Letters) 190, L13.
24. Epstein, R. I. and Petrosian, V., 1975, Ap. J. In press.
25. Field, G. B., 1974, Ap. J. 187, 453.
26. Field, G. B., 1975, Proceedings of Les Houches Summer School, ed. R. Ballian, P. Encrenax and J. Lequeux, to be published.

27. Fowler, W. A., Burbidge, G. R. and Burbidge, E. M., 1955, Ap. J. Auppl. 2, 167.
28. Geiss, J. and Reeves, H., 1972, Astr. Ap. 18, 126.
29. Gerola, H., Kafatos, M. and McCray, R., 1974, Ap. J. 189, 55.
30. Goret, P. and Beaudet, G., 1975, Preprint.
31. Gott III Jr., J. R., Gunn, J. E., Schramm, D. N. and Tinsley, B. M., 1974, Ap. J. 194, 543.
32. Hainebach, R. and Schramm, D. N., 1975a, Proc. 14th Int. Cosmic Ray Conf. 2, 549.
33. Hainebach, K., Schramm, D. N. and Blake, J. B. 1975b, submitted to Ap. J.
34. Hall, D. N. B. and Engvold, O., 1975, Ap. J. 197
35. Hoyle, F. and Fowler, W. A., 1973, Nature 241, 384.
36. Iben, I. J., 1973 in Explosive Nucleosynthesis, D. N. Schramm and W. D. Arnett ed., University of Texas Press p. 115.
37. Kirshner, R. P., Oke, J. B., Penston, M. V. and Searle, L. 1973, Ap. J. 185, 303.
38. Kozlovsky, B. and Ramaty, R., 1974, Ap. J. (Letters) 191, L43.
39. Meneguzzi, M., Audouze, J. and Reeves, H., 1971, Astr. Ap. 15, 337.
40. Meneguzzi, M. and Reeves, H., 1975, Astr. Ap. 40, 99.
41. Meszaros, P., 1974, Ap. J. 191, 79.
42. Milford, S. N. and Shen, B. S. P., 1961, Phys. Rev. 122, 1921.
43. Mitler, H. E., 1967, in High Energy Nuclear Reactions in Astrophysics, B. S. P. Shen, Benjamin, p. 59.
44. Mitler, H. E., 1972, Ap. and Space Sci. 17, 186.
45. Morton, D. C., 1974, Ap. J. (Letters) 193, L35.
46. Morton, D. C., Smith, A. M. and Stecher, J. P., 1974, Ap. J. (Letters) 189, L109.
47. Nogaard, H. and Arnould, M., 1975, Astron. Astroph. 40, 331.
48. Ostriker, J. P. and Tinsley, B. M., 1975, Ap. J. (Letters) in press.
49. Peebles, P. J. E., 1966, Phys. Rev. Letters 16, 410.
49a. Penzias, A. A., Wannier, P. G., Wilson, R. W. and Linke, R. A., 1976, Ap. J. (Letters) (submitted).
50. Ramadurai, S. and Wickramasinghe, N. C., 1975, Ap. and Space Sci. (Letters) 33, L41.
51. Reeves, H., 1971, Nuclear Reactions in Stellar Surfaces and Their Relations with Stellar Evolution (London: Gordon and Breach).
52. Reeves, H., 1972, Phys. Rev. D6, 3363.
53. Reeves, H., 1974a, Ann. Rev. Astr. Ap. 12, 437.
54. Reeves, H., 1974b, in Supernovae and Supernovae Remnants, Ed. C. B. Cosmovici, D. Reidel Co., p. 381.
55. Reeves, H., Audouze, J., Fowler, W. A. and Schramm, D. N., 1973, Ap. J. 179, 909.
56. Reeves, H., Fowler, W. A. and Hoyle, F., 1970, Nature 226, 727.
57. Rogerson, Jr., J. B. and York, D. G., 1973, Ap. J. (Letters) 186, L95.

58. Ryter, C., Reeves, H., Gradsztajn, E. and Audouze, J., 1970
    Astr. Ap. 8, 389.
59. Sackmann, I. J., Smith, R. L. and Despain, K. H., 1974,
    Ap. J. 187, 555.
60. Schramm, D. N. and Wagoner, R. V., 1974, Physics Today 27, 40.
61. Steigman, G., 1975, private communication.
62. Tinsley, B. M., 1974, Ap. J. 192, 629.
63. Toussaint, J., 1975, Thèse de 3e cycle, Université Paris XI
    unpublished.
64. Traub, W. A. and Carleton, N. P., 1973, Ap. J. (Letters)
    184, L11.
65. Truran, J. W., 1973, in Explosive Nucleosynthesis, ed. D. N.
    Schramm and W. D. Arnett. The University of Texas Press,
    p. 102.
66. Truran, J. W. and Cameron, A. G. W., 1971, Ap. and Space
    Sci. 14, 179.
67. Truran, J. W. and Cameron, A. G. W., 1975, in preparation.
68. Ulrich, R. K. and Scalo, J. M., 1972, Ap. J. 176, 137.
69. Van den Bout, P. A. and Grupsmith, J., 1973, BAAS 5, 380.
70. Vidal-Madjar, A., Laurent, C., Bonnet, R. M. and York, D. G.,
    1975, Proceedings of the XVIIIth COSPAR meeting, Verna,
    Bulgaria, June 1975, to be published.
71. Wagoner, R. V., 1973, Ap. J. 179, 343.
72. Wagoner, R. V., 1974, in Confrontations of Cosmological
    Theories, ed. M. S. Longair, D. Reidel Co., p. 195.
73. Wagoner, R. V., Fowler, W. A. and Hoyle, F., 1967, Ap. J.
    148, 3.
74. Weaver, T. A. and Chapline, G. F., 1974, Ap. J. (Letters)
    192, L57.
75. Yahil, A. and Beaudet, G., 1975, preprint.
76. York, D. G., 1975, Ap. J. (Letters) 196, L102.
77. York, D. G. and Rogerson, Jr., J. B., 1975, Ap. J., in press.
78. Zappala, R. R., 1972, Ap. J. 172, 57.

Note added in proof: Penzias and his collaborators (Penzias et al.
1976) have recently measured the ratio DCN/H ^{13}CN in several
molecular clouds. They performed the measurement for Sgr B2, for
which they found DCN/HCN $\simeq$ a few times 10^{-4}. This should correspond to D/H = $10^{-5.5\pm1}$. From this measurement Audouze et al.
(1976) concluded that the galactic center must have been very
recently replenished with fresh deuterium (the characteristic time
scale for astration in the galactic center is indeed as low as a
few times 10^8 years). They proposed either that some infall of
deuterium-rich matter may have occurred recently or that galactic
cosmic rays not only produce gamma rays but also D by spallation.

TABLE 1   COMPENDIUM OF THE RELEVANT OBSERVATIONS
(from Reeves, 1974)
The references for this table can be
in this review paper.

Table 1a.   Observations of deuterium

Observations	Technique	Relevant period	D/H in phase	Inferred cosmic D/H	Remarks
D in interstellar matter	Lyman lines	now	$1.4 \pm 0.2 \times 10^{-5}$	$1.4 \pm 0.2 \times 10^{-5}$	Uncertainties may be larger[a]
D in interstellar matter (toward galactic center)	96 cm	now	$<4 \times 10^{-4}$ $>3 \times 10^{-5}$	$<4 \times 10^{-4}$ $>2 \times 10^{-5}$	Lower limit could be lower[b]
D in interstellar matter (toward Cas A)	96 cm	now	$<7 \times 10^{-5}$	$<7 \times 10^{-5}$	-[c]
HD in clouds	Molecular absorption in UV	now	$\sim 10^{-6}$	$>5 \times 10^{-6}$ $<2 \times 10^{-4}$	-d,e
DCN in Orion nebula	Emission at 72 and 145 MHZ	now	up to 6×10^{-3} in HCN	10^{-6} to 10^{-5}	-f,g
HD in Jupiter	Absorption of molecular line at 7468 Å	Birth of Sun	$2.0 \pm 0.5 \times 10^{-5}$	$2.0 \pm 0.5 \times 10^{-5}$	Isotopic fractionation is likely to be small[h]

[a]Rogerson & York 1973.   [c]Weinreb 1962.   [e]Black & Dalgarno 1973.   [g]Solomon & Woolf 1973.

[b]Cesarsky et al. 1973.   [d]Spitzer et al 1973.   [f]Jefferts et al. 1973.   [h]Trauger et al. 1973.

Table 1a. (Continued)

Observations	Technique	Relevant period	D/H in phase	Inferred cosmic D/H	Remarks
CH_3D in Jupiter	Infrared absorption at 5μ	Birth of Sun	from 3 to 8×10^{-5} in methane	from 1 to 8×10^{-5}	Large uncertainties on effective equilibrium temperature[i,j,k]
D in protosolar nebula	Analysis of solar wind and water formation	Birth of Sun	16×10^{-5} in water	$2.5 \pm 1 \times 10^{-5}$	Uncertainties on formation T of water, [l,m]
D in protosolar nebula	Analysis of trapped He, Ne, A	Birth of Sun		$1.5 \, ^{+1.5}_{-0.7} \times 10^{-5}$	Uncertainties due to possible diffusion of He in meteorites[n]
D in solar surface	Photospheric lines	Birth of Sun	$<4 \times 10^{-6}$		D has been burned into ^3He[o]
D in solar surface	Coronal lines		$<4 \times 10^{-6}$		-p
D in solar surface	Studies of lunar dusts		$<3 \times 10^{-6}$		-q

[i]Beer et al. 1972. [j]Beer & Taylor 1973. [k]Reeves & Bottinga 1972. [l]Boato 1954.

[m]Geiss & Reeves 1972. [n]Black 1972. [o]Grevesse 1970a

[p]Hall, unpublished. [q]Epstein & Taylor 1972.

Table 1b. Lithium abundances

"Initial" abundances[a]

	Hyades	Praesepe	Pleiades	NGC 2264	T Tauri stars	FU Orionis
Li/H	8×10^{-10}	1.6×10^{-9}	10^{-9}	10^{-9}	5×10^{-10}	10^{-9}

"General" abundances

	Main sequence field stars	Red giants field stars	Solar	Chondrites
Li/H	up to 10^{-9}[b]	up to 10^{-7}[d,e]	$10-11$[f,g]	1.5×10^{-9}[h]
	$^6Li/^7Li \leqq 0.1$[c]		$^6Li/^7Li \leqq 0.1$[b]	$^7Li/^6Li = 12.5$[h]

Interstellar matter

Li/H	3×10^{-10}[i]
	6×10^{-10}[k]

For meteorites the normalization is made through the solar $Si/H = 3 \times 10^{-5}$[j]

[a] Zappala 1972.
[b] Wallerstein & Conti 1969.
[c] Cohen 1972.
[d] Boesgaard-Merchant 1970.
[e] Torres-Peimbert & Wallerstein 1966.
[f] Grevesse 1968.
[g] Engvold et al. 1970.
[h] Nichiporuk 1971.
[i] Traub & Carleton 1973.
[j] Cameron 1973.
[k] Van den Bout & Grupsmith 1973.

Table 1c. Beryllium abundances or upper limits

	Interstellar matter	Solar surface	Stars	Chondrites
Be/H	$<7 \times 10^{-11a}$	$10^{-11b,c,d}$ 2×10^{-10h}	4×10^{-11} to 10^{-12e}	2×10^{-11f}

For meteorites the normalization is made through the solar Si/H = 3×10^{-5g}

[a]Boesgaard-Merchant 1974.

[b]Grevesse 1968.

[c]Aller & Ross 1974.

[d]Hauge & Engvold 1968.

[e]Wallerstein & Conti 1969.

[f]Buseck 1971.

[g]Cameron 1973.

[h]Shipman 1974.

Table 1d. Boron abundances or upper limits

	Interstellar matter	Solar surface	Carbonaceous chondrites	Enstatite chondrites
B/H	$<2 \times 10^{-9a,b}$	$<6 \times 10^{-10c}$	$3 \text{ to } 6 \times 10^{-9}$	$2 \text{ to } 5 \times 10^{-10}$
			$^{11}B/^{10}B = 4^d$	$^{11}B/^{10}B \approx 4^d$

For meteorites the normalization is made through the solar Si/H = 3×10^{-5e}

[a]Morton et al. 1973.

[b]Audouze et al. 1973.

[c]Grevesse 1968.

[d]Baedecker 1971.

[e]Cameron 1973.

TABLE 2   THE "BEST CHOICE" FOR COSMIC ABUNDANCES OF LIGHT ELEMENTS

$$\frac{Li}{H} = 10^{-9 \pm 0.3}$$

$$\frac{Be}{H} = 10^{-10.9 \pm 0.3}$$

$$\frac{B}{H} = 10^{-10 \pm 0.3}$$

$$\frac{^{7}Li}{^{6}Li} = 12.5 \pm 0.5$$

$$\frac{^{11}B}{^{10}B} = 4 \pm 0.4$$

$$\frac{B}{Be} = 10^{0.9 \pm 0.3}$$

TABLE 3   FORMATION RATE OF LIGHT ELEMENTS THROUGH THE
BOMBARDMENT OF THE INTERSTELLAR MATTER BY
THE GALACTIC COSMIC RAYS[*]

Light Elements (1)	GCR Production Rate $(g^{-1} s^{-1})$ (2)	Abundance due to GCR (relative to H) (3)
^{6}Li.........	1.1×10^{-4}	8×10^{-11}
^{7}Li.........	1.7×10^{-4}	1.2×10^{-10}
^{9}Be.........	2.8×10^{-5}	2×10^{-11}
^{10}B.........	1.2×10^{-4}	8.7×10^{-11}
^{11}B.........	2.8×10^{-4}	2×10^{-10}

[*]From Meneguzzi, Audouze, and Reeves 1971.

NOTE.--Column (2) gives the number of atoms per gram of
interstellar matter (of standard cosmic composition) per
second. Column (3) gives the ratio of hydrogen (per
number of atoms) after 10^{10} years of bombardment at the
present rate (neglecting stellar destruction).

NUCLEON-MESON TRANSPORT CALCULATIONS*

R. G. Alsmiller, Jr.

Neutron Physics Division, Oak Ridge National
Laboratory, Oak Ridge, Tennessee, 37830

ABSTRACT. A review of medium- and high-energy nucleon-meson
transport calculations carried out at the Oak Ridge National
Laboratory in the past several years is presented.

1. INTRODUCTION

For a variety of applications -- for example, the study of the
shielding required for radiation protection in the vicinity of
medium- and high-energy accelerators, the design of instrumentation
for measuring the energy of high-energy particles, the study of the
induced activity in the moon from solar and galactic cosmic-ray
bombardment, and the study of the applicability of various particles in cancer radiotherapy -- it is desirable to have available
calculational methods for studying the transport of medium- and
high-energy nucleons and pions through matter. The equations which
describe the transport of these particles and the nuclear-reaction
products produced by these particles through matter are well established, but the problem of obtaining accurate numerical solutions
to the equations for specific conditions is quite formidable. Also, the basic differential particle-production cross-section data
that enter into the equations are not available experimentally,
and, thus, nucleon- and pion-transport calculations must be based,
for the most part, on cross-section data derived from theoretical
considerations. For several years, a group at the Oak Ridge

---

* This research was funded by the U. S. Energy Research and
  Development Administration under contract with the Union
  Carbide Corporation.

National Laboratory have been engaged in developing nuclear cross-section data and techniques for carrying out medium- and high-energy transport calculations, and in this paper some of the results which have been obtained are reviewed.

2.   HETC - HIGH-ENERGY TRANSPORT CODE

HETC[1]* is a transport code developed at ORNL that utilizes Monte Carlo techniques to simulate the nucleon-meson cascade that occurs when high-energy nucleons or pions pass through matter. A detailed discussion of the physical processes included in this code is beyond the scope of this paper, but such a discussion is available.[3,4] Here, only a few of the more significant aspects of the code will be discussed.

HETC is operable for incident neutrons, protons, charged pions, and muons at all energies less than or of the order of several hundred GeV and for essentially arbitrary material compositions and source-geometry configurations. The code takes into account charged-particle energy loss due to atomic ionization and excitation, multiple Coulomb scattering of incident charged particles, elastic and nonelastic neutron-nucleus collisions, nonelastic collisions of protons and charged pions with nuclei other than hydrogen, elastic and nonelastic collisions of nucleons and charged pions with hydrogen, pion and muon decay in flight and at rest, and particle production from negatively charged pion capture at rest. At energies $\leqslant$ 15 MeV, neutron transport in HETC is usually carried out with the Monte Carlo transport code 05R[5] using a combination of experimental[6,7] and theoretical cross-section data.[8] For some purposes, however, it is more convenient to transport the low-energy neutrons by the method of discrete ordinates, and this may also be done.[9] At energies $\geqslant$ 15 MeV, nucleon-hydrogen and pion-hydrogen collisions are treated using basically experimental data,[10-13] and neutron-nucleus elastic collisions are treated using experimental data and optical-model calculations (e.g., see Refs. 14 and 15 and the many references given therein).

For both nucleons and pions, there is little information on particle production from nonelastic nuclear collisions with nuclei heavier than hydrogen, and thus this information must be obtained theoretically. In HETC, nucleon-nucleus nonelastic collisions in

---

* HETC is a high-energy version of an earlier code designated NMTC.[2] For incident nucleon and pion energies $\leqslant$ 3 GeV, there is no significant difference between HETC and NMTC. HETC and an instruction manual for operating the code are available from the Radiation Shielding Information Center of the Oak Ridge National Laboratory.

the energy range of 15 MeV to 3.5 GeV and pion-nucleus nonelastic collisions at energies < 2.5 GeV, including negatively charged pion capture at rest, are treated by means of the intranuclear-cascade-evaporation model of nuclear reactions as implemented by Bertini *et al.* in the code MECC-7.[16]* A detailed discussion of this model and comparisons between calculated results and a variety of experimental data will be found in Refs. 11 and 17-21. This model was intended to describe nonelastic nuclear collisions at energies $\gtrsim$ 50 MeV, but because of the lack of any other source of data, it is used in HETC at lower energies despite the fact that its use at these lower energies is dubious. Comparisons between calculated and experimental data at these lower energies are given in Refs. 22-25. In particular, the case of particle production from negatively charged pion capture at rest, which is of importance in the study of the applicability of negatively charged pions in radiotherapy, is considered in some detail in the work of Guthrie *et al.*[26] A unique feature of HETC is that it contains the code MECC-7 as a subroutine and thus uses the code itself rather than data from the code to determine the collision products when a nonelastic collision occurs. This use of MECC-7 as a subroutine substantially reduces the data-handling problems in HETC, but it also makes the core requirements for HETC very large since MECC-7 is not a small code.

The intranuclear-cascade-evaporation model as programmed in MECC-7, and therefore in HETC, becomes invalid at nucleon energies ~ 3.5 GeV and pion energies ~ 2.5 GeV because the model accounts for only double-pion production from nucleon-nucleon collisions and for only single-pion production from pion-nucleon collisions. At nucleon energies > 3.5 GeV and pion energies > 2.5 GeV, particle-production data for nonelastic collisions are obtained in HETC from a very approximate extrapolation model due to Gabriel *et al.*[27] and to Gabriel and Santoro.[28] A detailed discussion of this extrapolation model, as well as a large number of comparisons between calculated and experimental data, will be found in Ref. 27 and in Appendix A of Ref. 28. Basically, the model employs particle production data obtained from the intranuclear-cascade-evaporation model at energies ~ 3 GeV and uses energy- and angle-scaling relations to estimate particle-production data for higher energy nucleon nucleus and pion-nucleus collisions. In HETC the intranuclear-cascade-evaporation data, needed for scaling, are generated by the code MECC-7 at each nonelastic collision, and, therefore, as in the intermediate energy range, it is not necessary to store large

---

* Except for improved nucleon-nucleon and pion-nucleon cross-section data at the higher energies (> 350 MeV), MECC-7 is the same as MECC-3[16] and is a medium-energy extension of an earlier code designated LECC.

quantities of data which describe particle production from high-
energy nucleon-nucleus and pion-nucleus collisions. The extrapolation model is at best very approximate, but it does provide a
detailed description of each high-energy collision; e.g., it gives
the mass, charge, and energy of the residual nucleus from the collision, and for this reason it is very suitable for use in transport calculations.

In HETC, all particles in the cascade, i.e., neutrons, pro-
tons, π^+, π^-, μ^+, and μ^-, are followed until they eventually disappear by escaping from the geometric boundaries of the system,
undergo nuclear absorption, come to rest, or, in the case of pions
and muons, decay. Electrons and positrons from muon decay and
photons from $\pi^°$ decay and nonelastic nuclear collisions are not
transported by HETC, but information concerning these particles
is provided by the code, so that they may be transported with
other codes if this is an essential part of the problem. (In the
case of photons from nonelastic nuclear collisions, the code gives
the total energy emitted from the collision in the form of photons
but not the photon spectrum.) HETC also provides the mass number,
charge, and location of all of the residual nuclei produced by
nonelastic collisions. This information allows a determination of
the induced activity in the transport medium as a result of the
cascade and, for this reason, is of interest in assessing the
radiation hazard in the vicinity of accelerators and in studies of
the effects of cosmic-ray bombardment. Finally, HETC provides the
energy distribution of all of the residual nuclei produced by nuclear collision and the energy distribution of the low-energy heavy
particles, i.e., deuterons, tritons, ^{3}He's, and alpha particles,
produced by nuclear reactions. These energy distributions are
very important in calculating the dose equivalent for radiation-
protection purposes and in calculating quantities of biomedical
interest such as cell-survival probabilities. In the next section
of this paper, examples of these various types of calculations are
presented.

3.   CALCULATED RESULTS AND COMPARISONS WITH EXPERIMENTAL DATA

In this section, calculated results pertaining to a variety
of applications are presented and discussed. For the most part,
the applications considered here are those for which experimental
data are available, since the validity of the calculated results
can be determined only by comparison with such data. The value
of such calculations, however, is not only that they may be used
to understand existing experimental data, but also that they provide results which are not available and cannot easily be obtained
experimentally. All of the calculated results presented in this
section, unless otherwise noted, were obtained with the code HETC
described in the previous section.

3.1 Neutron flux spectra induced in the earth's atmosphere by galactic cosmic rays[9]

The galactic cosmic-ray proton spectrum contains very high energy protons, and, therefore, the nucleon-meson cascade induced in matter by galactic cosmic rays is of considerable interest for testing the validity of high-energy transport calculations. Furthermore, the neutron flux per unit energy induced in the earth's atmosphere by galactic cosmic rays is of interest for a variety of reasons: e.g., one theory of the origin of the Van Allen proton belt is based on the existence of this neutron flux at the top of the atmosphere,[29] and, therefore, experimental data are available on this neutron flux per unit energy as a function of both neutron energy and depth in the atmosphere.[30-36]

Armstrong et $al.$[9] have compared calculated results, obtained with both incident protons and incident alpha particles, with the available experimental data. The calculations for incident alpha particles were carried out using a modified version of HETC, which employs an approximate model due to Gabriel et $al.$[37] to describe particle production from high-energy nonelastic alpha-particle-nucleus collisions. In the calculations, the low-energy (< 12 MeV) neutrons were transported with the discrete ordinates code ANISN[38] rather than with the Monte Carlo code O5R[5] usually used in HETC in order that the energy dependence of the thermal neutron flux could be obtained. The calculations took into account the density and temperature variation of the atmosphere as a function of depth and the presence of the ground. The atmosphere was assumed to be composed of 79% nitrogen and 21% oxygen and was approximated by a semi-infinite slab 1033-g-cm^{-2} thick. The ground was also approximated by a slab and was assumed to be SiO_2. The galactic cosmic-ray proton and alpha-particle spectra were taken from the review article of McDonald,[39] and the effect of the geomagnetic fields at all angles of incidence was taken into account.[40]

In Fig. 1 the calculated results are compared with the experimental data of Hess et $al.$[32] The error bars on the calculated results, where shown, are statistical and represent one standard deviation. At the lower neutron energies (< 12 MeV), error bars are not given because the low-energy neutron transport calculations were not carried out with Monte Carlo techniques. At depths of 200 g cm^{-2} and 1033 g cm^{-2}, calculated results are shown for incident protons and for both incident protons and alpha particles to give an estimate of the effect of the alpha particles. The measured and calculated spectra differ somewhat at low energies, especially in the thermal-energy region, but are in quite good agreement at the higher energies. The accuracy of the measured spectra at very low energies ($\leqslant 4 \times 10^{-8}$ MeV) is estimated by Hess et $al.$[32] to be about a factor of 2. In considering the

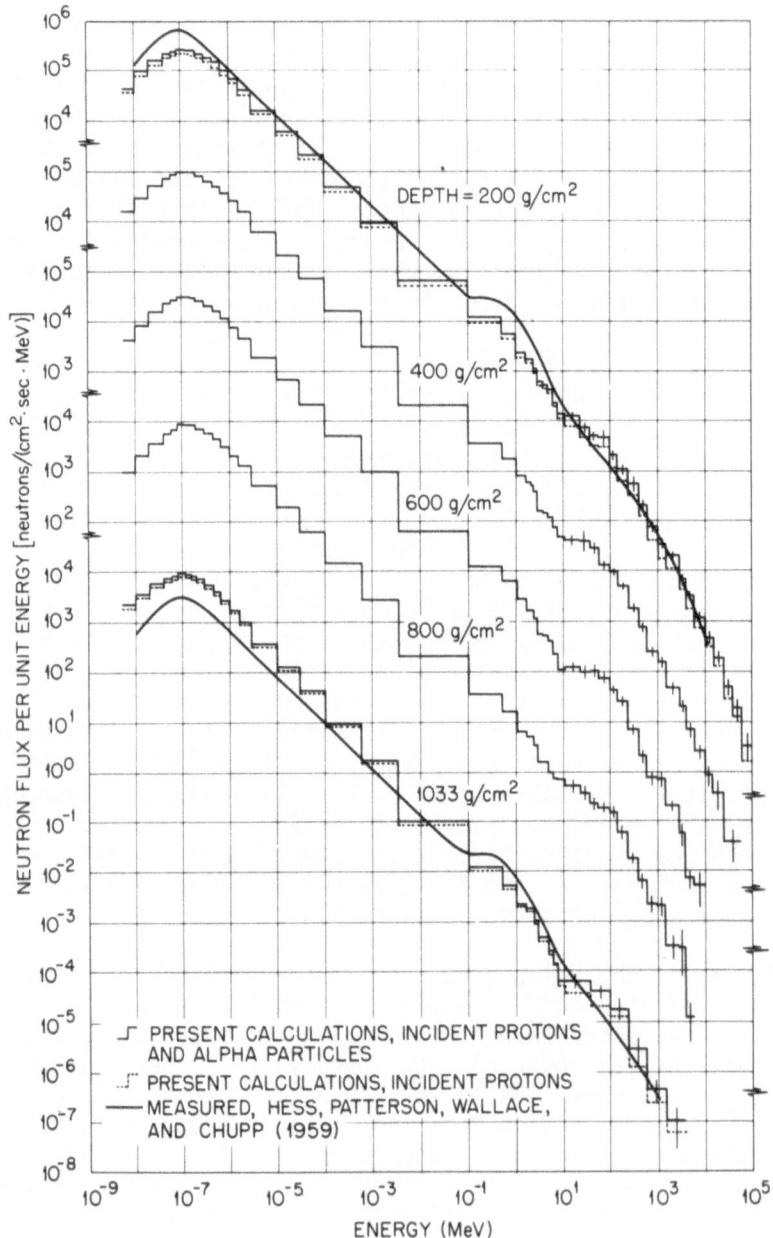

Fig. 1. Omnidirectional (4π) neutron flux per unit energy
at various depths from the top of the atmosphere (solar
minimum, geomagnetic latitude = 42°N).

results, it must be remembered that the calculated results are absolute and have not in any way been normalized to the experimental data.

Recently Preszler *et al.*[36] made detailed measurements on the omnidirectional neutron flux per unit energy near the top of the atmosphere in the energy range of 10 to 100 MeV. In Fig. 2 the calculated results of Armstrong *et al.*[9] are compared with these data.* Also shown are the earlier data of Hess *et al.*[32] at a depth of 200 g cm^{-2}. The agreement between the calculated results and the experimental data of Preszler *et al.* is very good at all energies and depths considered.

3.2 Lateral development of the nucleon-meson cascade
     induced in iron by 29.4-GeV/c protons[4]

In section 3.1, the longitudinal development of a nucleon-meson cascade was considered; that is, the properties of the cascade changed as a function of depth in the medium, but, because of the broad spatial extent of the incident proton and alpha-particle beams, the properties of the cascade could be assumed to be independent of position coordinates measured transversely to the depth coordinate. The lateral development of the nucleon-meson cascade induced in matter by a narrow beam of incident particles, which is of importance in some accelerator shielding applications, will now be considered. The lateral development of a cascade is much more dependent on the angular distribution of particles produced by nonelastic nuclear collisions than is the longitudinal development of a cascade, and, thus, the comparisons between calculated and experimental results presented here provide a more stringent test of the calculations than those presented in 3.1.

Van Ginneken and Borak[42] have made measurements on the lateral development of the cascade induced in a very thick iron shield by a narrow beam of 29.4-GeV/c protons. The quantity measured was the production of ^{18}F in aluminum foils placed in the iron at various depths and at various distances from the axis of the incident proton beam.

Armstrong *et al.*[4] carried out calculations with HETC and made comparisons between the calculated results and the experimental

---

* Comparisons between calculated results and the experimental data of Preszler *et al.* have also been given by Merker.[41]

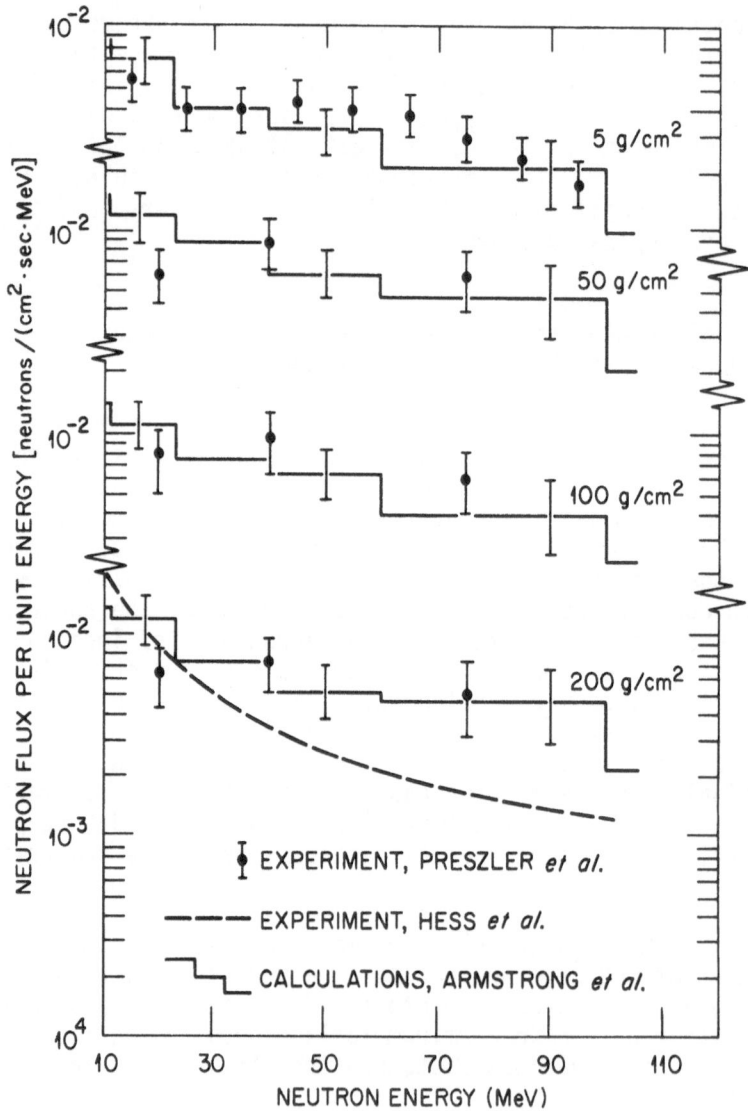

Fig. 2. Omnidirectional (4π) neutron flux per unit energy at various depths from the top of the atmosphere. (This figure is reprinted from Ref. 36 with permission.)

data of Van Ginneken and Borak.* In the experiment, the target
was made up of slabs of iron of various thicknesses with air gaps
between the slabs. These air gaps were taken into account in the
calculations. The calculations were performed by first determining the energy and spatial distributions of the nucleon and
charged-pion fluxes in the experimental configuration and then determining the ^{18}F production by using these fluxes in conjunction
with the energy-dependent cross sections for the production of
^{18}F from aluminum by the various particles.[44] The energy dependence of these cross sections is such that the ^{18}F production may
be interpreted as a measure of the nucleon and charged-pion fluxes
above ~ 50 MeV.

Comparisons between the calculated and experimental results
are shown in Figs. 3 (a), (b), and (c). The comparisons are absolute and unnormalized. The agreement is quite good although
the calculations give a consistently higher production near the
beam axis. Because of the small area and alignment of the aluminum foils used in the experiment, however, the experimental ^{18}F
production is expected to yield an underestimate of the actual
production very near the beam axis.[45]

3.3 Energy-deposition fluctuations in an iron-liquid
     scintillator ionization spectrometer[46]

In section 3.2, the average properties of a nucleon-meson
cascade were considered; i.e., the results presented were obtained
by averaging over a large number of incident particles. Because
of the statistical nature of the physical processes involved in
the cascade, each incident particle of a given type and energy
does not produce identical physical results, and for some applications the variation in the results produced by individual incident particles is of considerable importance. One such application which will now be considered is the performance of ionization
spectrometers.

Basically, an ionization spectrometer is a collection of metal
slabs interspersed with liquid scintillator slabs, which is used
to measure the energy of high-energy protons and pions by measuring the light output from the liquid scintillators. The ability
of the spectrometer to give an estimate of the energy of an incident particle of a given type is dependent on there being a unique
relationship between the energy of the incident particle and the

---

* Calculations have also been made for this experimental configuration by Van Ginneken and Borak[42] using the code FLUTRA which
  is a modified version of a Monte Carlo code due to Ranft.[43]

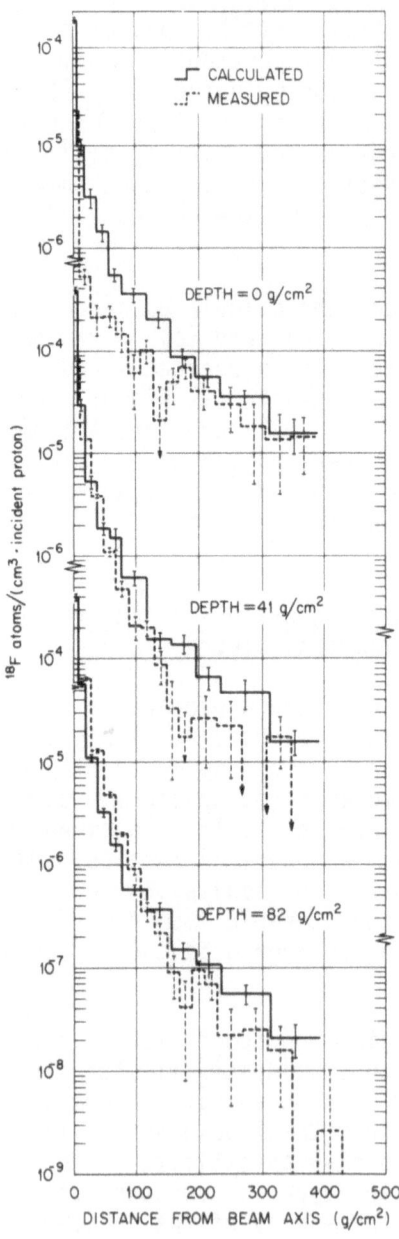

Fig. 3 (a). Comparison of Calculated and experimental[42] lateral distributions of ^{18}F production in aluminum foils at various depths in iron for 29.4-GeV/c incident protons.

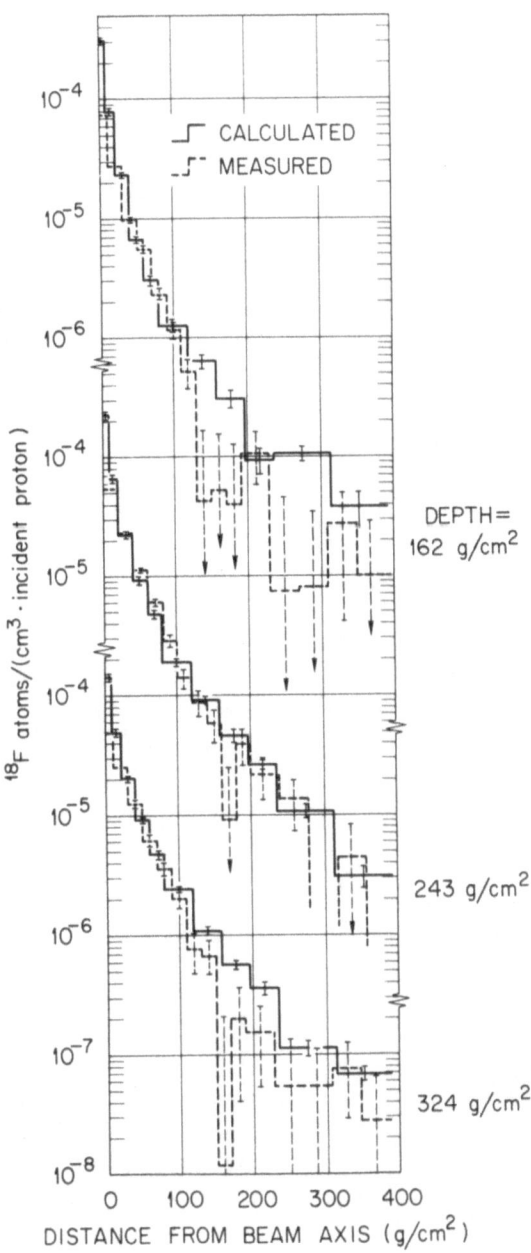

Fig. 3 (b). Comparison of Calculated and experimental[42] lateral distributions of ^{18}F production in aluminum foils at various depths in iron for 29.4-GeV/c incident protons.

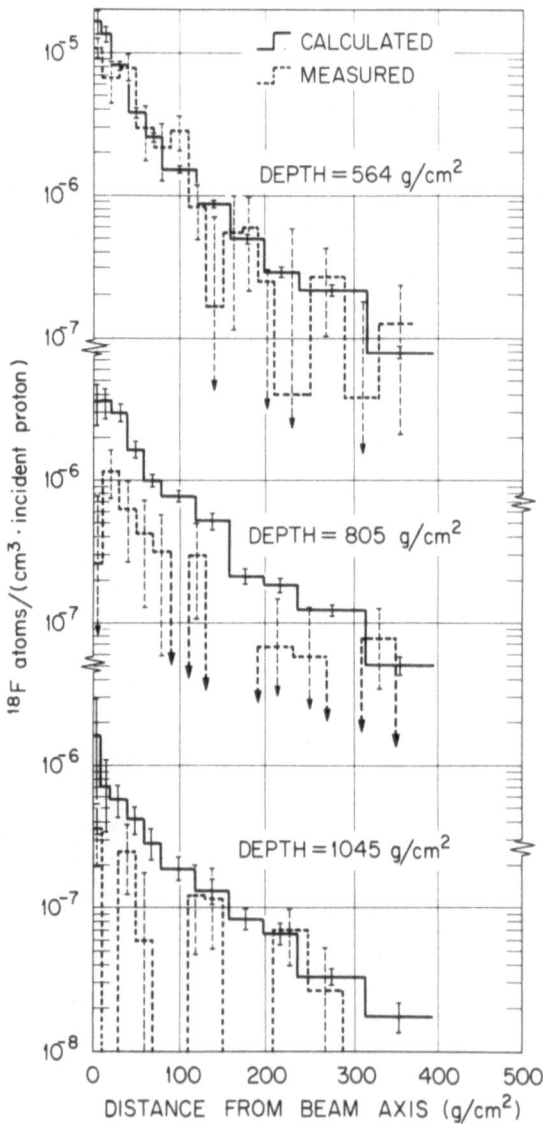

Fig. 3 (c). Comparison of calculated and experimental[42] lateral distributions of ^{18}F production in aluminum foils at various depths in iron for 29.4-GeV/c incident protons.

light output of the liquid scintillators. Such a unique relationship does not exist because of the variations in the energy deposition in the scintillator from individual incident particles of
the same type, energy, direction, and initial coordinates. The
validity of the energy estimate obtained from the spectrometer is,
therefore, very dependent on the magnitude of the energy-deposition
fluctuations, and thus a knowledge of the magnitude of these fluctuations is of considerable importance.

   In Table I the composition and dimensions of the spectrometer
considered are given. This spectrometer design and the experimental data pertaining to this design are due to Selove and his
collaborators.[47] Gabriel et al.[46] have carried out calculations
with HETC for this design and have made comparisons between the
calculated and experimental data. For this application, the
transport of the high-energy photons from π° decay and the transport of the electrons and positrons from muon decay cannot be
neglected. The code HETC does not transport these particles, but
it does produce a source distribution, i.e., the energy, direction,
and position of each type of particle, so they may be transported
with another code. In obtaining the results presented here, the
electron-photon cascade code of Beck[48] was used to transport the
electrons, positrons, and photons produced by the cascade. In the
calculations, the nonlinearity of the light pulse from the scintillators (i.e., the fact that the light observed is not directly
proportional to the energy deposited) was taken into account by
the use of Birks' law.[49] In the calculations and insofar as possible in the experiment, the incident particles were normally incident at the geometric center of the iron plate in front of the
first cell.

   In Figs. 4 and 5 calculated and experimental results for
15.0-GeV/c incident protons and positively charged pions, respectively, are presented and compared. The data shown indicate the
frequency with which a given scintillator pulse height is produced
as a function of pulse height. The calculated and experimental
histograms in each case are normalized to the same number of incident particles in the pulse-height region over which experimental
data are given. The error bars on the calculated histograms are
statistical only and represent one standard deviation. In both
cases, the calculated and experimental results are in good agreement, but in the case of incident pions, the calculated results
are slightly larger than the experimental results for pulse heights
of ~ 4500 MeV.

   Additional results on calculated energy-deposition fluctuations and the performance of ionization spectrometers have been
given by Gabriel and Chandler[50] and by Gabriel and Amburgey.[51]

Table I

Spectrometer Design

Basic Cell

Thickness (cm)	Material
0.32	Fe
3.81	Liquid Scintillator[a]
1.27	Fe
3.81	Liquid Scintillator[a]
1.27	Fe
3.81	Liquid Scintillator[a]
0.32	Fe

The spectrometer is composed of 15 basic cells. In front of the first cell, there is 0.95 cm of Fe and between each cell there are

0.71 cm	Void
1.91 cm	Lucite[b]
0.71 cm	Void

The lateral dimensions of the spectrometer are:

122 cm^2 by 122 cm^2

a. CH_2, $\rho = 0.87$ g cm^{-3}

b. $C_5H_8O_2$, $\rho = 0.944$ g cm^{-3}

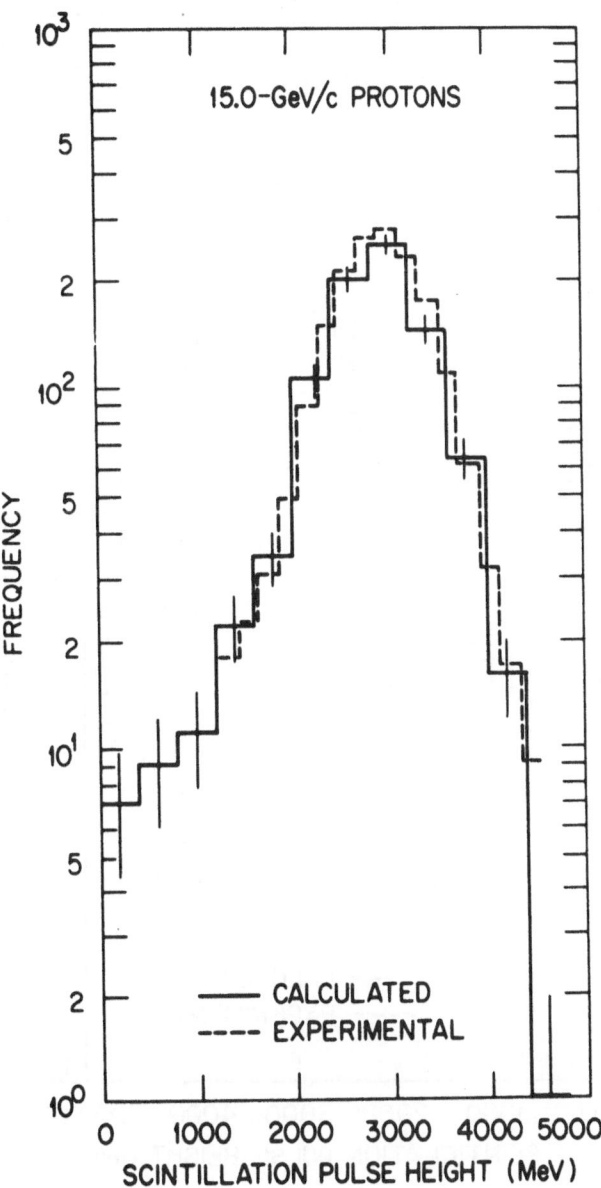

Fig. 4. Calculated and experimental pulse-height distribution for 15.0-GeV/c protons incident on a particular ionization spectrometer (see Table I).

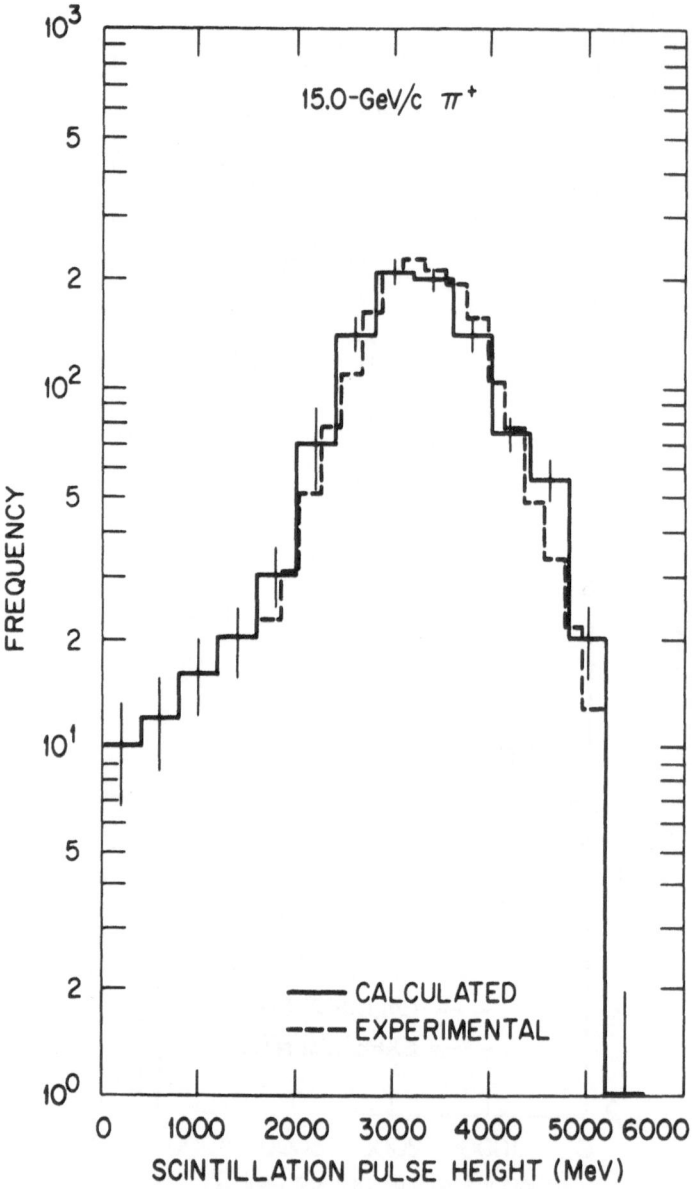

Fig. 5. Calculated and experimental pulse-height
distribution for 15.0-GeV/c positively charged pions
incident on a particular ionization spectrometer
(see Table I).

3.4 Induced activity in the moon from solar and galactic cosmic-ray bombardment[52]

When a high-energy nucleon-meson cascade takes place in matter, a variety of residual nuclei are produced as a result of non-elastic nuclear collisions. The production of these residual nuclei is important in accelerator shielding as well as in a variety of other fields.[53,54] In the case of accelerators, some of the radioactive residual nuclei produced in the cascade emit photons and thus present a potential radiation hazard after the accelerator has been turned off. Here, calculated results of the induced activity in the moon from solar and galactic cosmic-ray bombardment are presented as an example of induced-activity calculations performed with HETC. This example is chosen because of the availability of experimental data with which to compare the calculated results. The calculational method employed is, in principle, no different from that which has been used in calculating the induced activity in the vicinity of high-energy accelerators.[55-57]

As a result of the Apollo flights to the moon, experimental data on several radionuclides as a function of depth in the moon are available. Armstrong and Alsmiller[52] have carried out calculations and made comparisons with these experimental data. The lunar composition and density used in the calculations were taken from Apollo 11 measurements for type D fine material.[58] The calculations were carried out for an isotropic flux of galactic or solar protons incident on a half-space of lunar material. The galactic solar-minimum proton spectrum was taken from the review article of McDonald[39] and the galactic solar-maximum proton spectrum was taken from the wind-modulation theory of Durgaprasad *et al.*[59] The solar cosmic-ray proton data for solar cycle 19 were taken from the work of Modisette *et al.*[60] and of Weber,[61] and for solar cycle 20 through November 1969 were taken from the work of Hsieh,[62] Yucker,[63] Radke,[64] Baker,[65] and the compilation of the ESSA Research Laboratories.[66] In the calculations, it was assumed that the data for solar cycle 19 could be applied to all previous solar cycles. The solar and galactic cosmic-ray proton data used and a discussion of the manner in which the time dependence of the production and decay was taken into account are given in Ref. 67.

In Figs. 6 (a) and 7 (a), the calculated ^{26}Al and ^{22}Na activity, respectively, from galactic cosmic-ray protons at solar maximum and solar minimum and from solar cosmic-ray protons is shown as a function of depth. In Fig. 6 (b), the total ^{26}Al activity, i.e., the sum of the solar contribution and the average of the galactic contributions at solar maximum and solar minimum, is compared with the activity measured by Shedlovsky *et al.*[68] and by Finkel *et al.*[69] In Fig. 7 (b), the total ^{22}Na activity is compared with the activity measured by Shedlovsky *et al.*[68] In all of the figures, the activity is expressed in disintegrations per

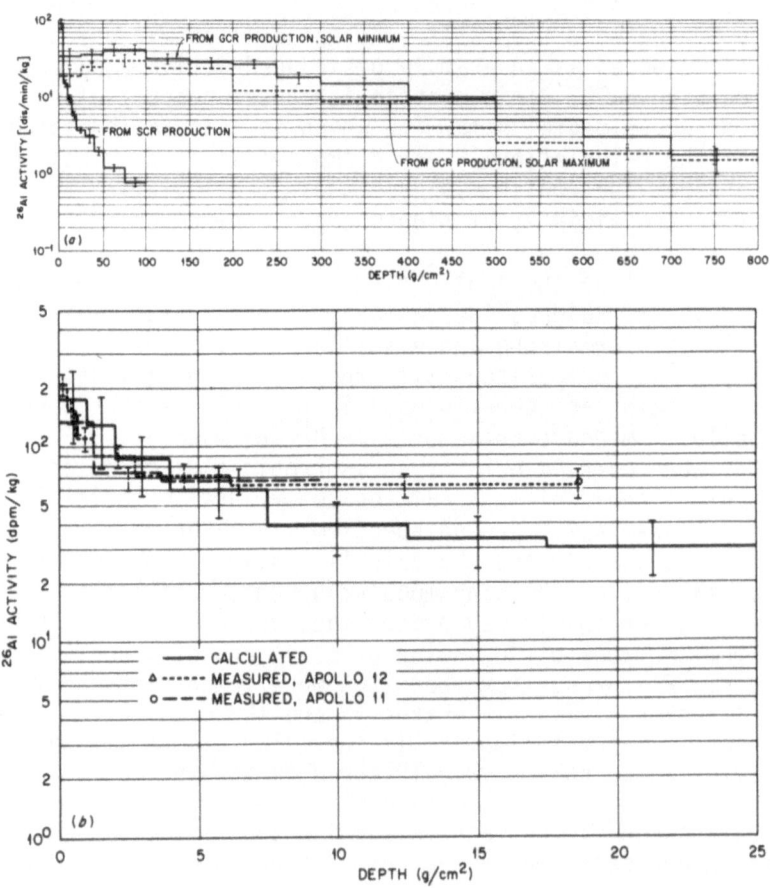

Fig. 6. (a) ^{26}Al activity vs depth for incident solar and
galactic cosmic-ray protons; (b) comparison of calculated
and measured ^{26}Al activity as a function of depth. (The
Apollo 11 data are from Shedlovsky *et al.*[68] and the Apollo 12
data are from Finkel *et al.*[69])

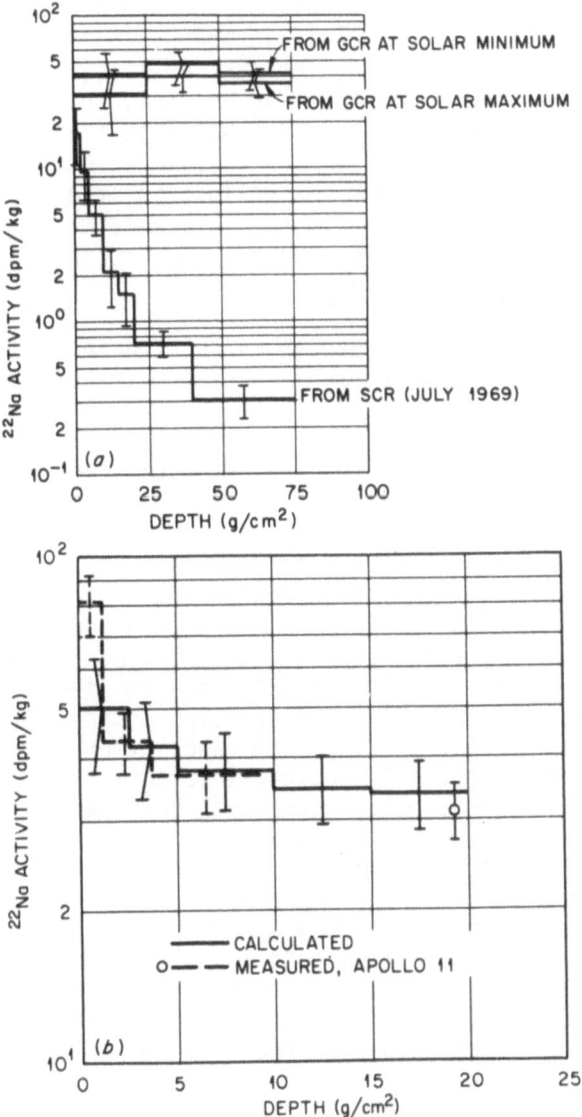

Fig. 7. (a) Depth dependence of ^{22}Na activity from solar proton bombardment at the time of the Apollo 11 flight and from galactic proton bombardment; (b) comparison of calculated and measured ^{22}Na activity as a function of depth. (The Apollo 11 data are from Shedlovsky *et al.*[68])

minute per kilogram. The ^{26}Al activity is nearly constant over a
time interval of one solar cycle because of the long ^{26}Al half-
life (7.4×10^5 y), but the ^{22}Na activity varies widely over a
solar cycle because of the relatively short ^{22}Na half-life (2.62 y).
The comparisons between the calculated and experimental data in
Figs. 6 (b) and 7 (b) are absolute in the sense that there are no
free parameters in the theory. The ^{26}Al activity for depths of
$\leqslant 10$ g cm^{-2} is due primarily to solar protons, whereas the activity at greater depths is due primarily to galactic protons. Therefore, the good agreement between the calculated and experimental
results in Fig. 6 (b) confirms the validity of the calculations
for both solar and galactic protons. In general, the agreement
between the calculated and experimental results in both Figs. 6 (b)
and 7 (b) is quite satisfactory.

Additional comparisons between calculated induced activity
and experimental data have been given by Armstrong.[70]

3.5 Absorbed doses and cell-survival probabilities from negatively charged pions[71,72]

The code HETC has been used to calculate the energy deposition in tissue by a variety of charged-pion, neutron, and proton
beams,[71-77] and results obtained with the code have, in general,
been found to be in good agreement with experimental data.[71,76,77]
The usefulness of the code is not only that it gives a reliable
estimate of the absorbed dose as a function of position in a phantom but also that it gives a complete description of all of the
charged particles produced by nuclear interactions, and this information may be used to obtain the dose equivalent for radiation-
protection purposes or may be combined with cell-inactivation
models[78,79] to obtain quantities of biomedical interest, such as
cell-survival probabilities. As an example of these types of calculations, results are presented and compared with experimental
data for the absorbed dose as a function of depth in a tissue
phantom irradiated by a negatively charged pion beam and for the
cell-survival probability as a function of absorbed dose in the
capture region of a negatively charged pion beam.

Experimental data on the absorbed dose as a function of depth
in a water phantom irradiated by a negatively charged pion beam
with a mean momentum of 175 MeV/c (corresponding to a kinetic energy of 84 MeV) are available from Turner et al.[80] In the experiment, the pion beam was approximately elliptical in shape and was
normally incident on one end of a water phantom. The absorbed
dose was measured as a function of depth along the beam axis using
a detector 3 cm in radius. The momentum spread of the beam was
not measured precisely but was estimated to be Gaussian with a
standard deviation of 4.7 MeV/c. The experimental π^- beam was

contaminated with unmeasured numbers of negatively charged muons and electrons, and, thus, the measured absorbed-dose data contain contributions from these particles as well as from pions.

Using HETC for π^- and μ^- transport and the electron-photon cascade code of Beck[48] for e^- transport, calculations have been carried out by Armstrong and Chandler[71] for comparisons with the experimental data.* In order to compare with the data, separate dose distributions for incident π^-, μ^-, and e^- beams were obtained, and the relative contribution of each type of particle to the measured absorbed-dose curve was determined by fitting the calculated results to the measured curve at three points. The absorbed dose is shown in Fig. 8 as a function of depth from the three different incident beams, and the comparison between the calculated and experimental data is shown in Fig. 9. Depths of 20.5, 28.5, and 35 cm were used to determine the relative number of each type of incident particle, so the calculated and experimental results must agree at these points. The calculations indicated that the beam was composed of 71% π^-, 13% μ^-, and 16% e^-. The agreement between the calculated and experimental data in Fig. 9 is quite satisfactory, but the comparison is not definitive because of the fitting procedure needed to determine the beam contamination.

Katz *et al.* have developed a model of cell inactivation.[78,79] Basically, this model allows a prediction of the probability that a given cell type will survive in any known radiation field no matter how complex, provided that the parameters in the model are available. These parameters, which must be determined from experimental data, are dependent on cell type and on whether the irradiation is carried out under aerobic or anoxic conditions. Parameters are presently available for a few cell types. The Katz *et al.* model, in conjunction with the data obtained from HETC, may be used to calculate cell-survival probabilities as a function of depth in a tissue phantom. A detailed discussion of the manner in which such calculations may be carried out has been given by Armstrong and Chandler.[72]

Experimental data on the spatial dependence of cell-survival probabilities when various particles are incident on a phantom are not available at present for comparison with calculated results. There is, however, one experiment by Raju *et al.*[82] in which the survival of T-1 kidney cells was measured at a single location in

---

* Comparisons between calculated results, obtained using a different calculational procedure than that used by Armstrong and Chandler, and these experimental data have also been given by Turner *et al.*[81]

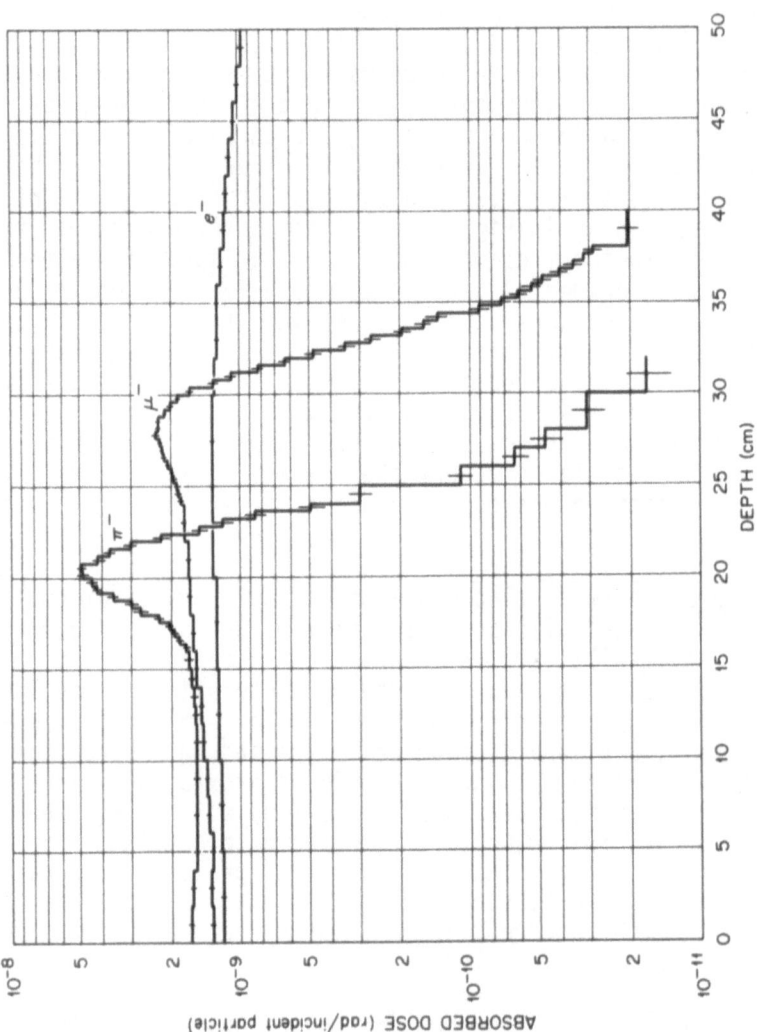

Fig. 8. Absorbed dose vs depth along the beam center line for π^-, μ^-, and e^- beams with a mean momentum of 175 MeV/c incident on a water phantom.

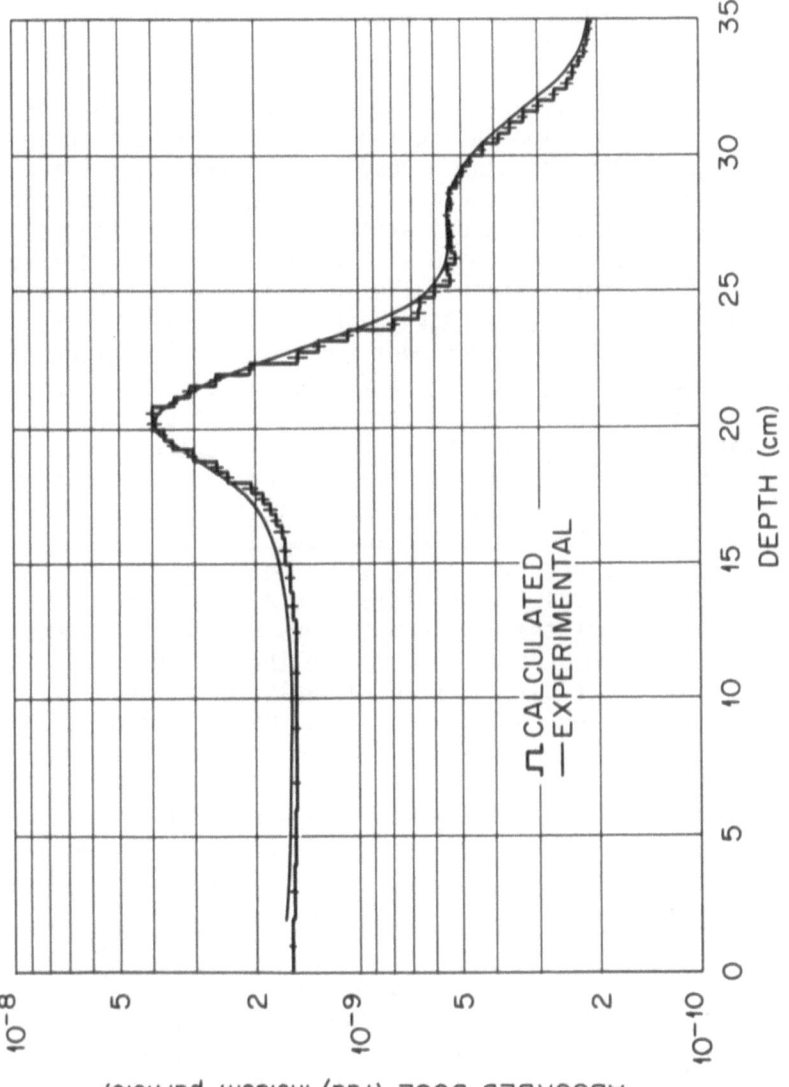

Fig. 9. Comparison of calculated and experimental absorbed dose from an incident π^- beam.

the stopping region of a pion beam. The measurements were made
by placing T-1 kidney cells in the peak-dose region of a contaminated pion beam (65% π^-, 10% μ^-, and 25% e^-) having a mean range
of ~ 23 g cm^{-2} in lucite. Experimental cell-survival probabilities taken under both anoxic and aerobic conditions are shown in
Fig. 10 as a function of absorbed dose. The experimental data
show considerable spread because the experiment had to be performed with a very low intensity beam.

     Also shown in Fig. 10 are calculated results due to Armstrong
and Chandler.[72] Calculations using the experimental beam could
not be made because the momentum spread of the beam was not measured. The results of Armstrong and Chandler were obtained using
a pion beam which produced approximately uniform pion captures
over the depth interval of 12.5 to 17.5 cm and the radial interval
of 0 to 2.5 cm in tissue. The results shown in the figure are the
cell-survival probabilities averaged over this capture region.
While the conditions in the experiment and in the calculations are
not completely equivalent, they are sufficiently similar to permit
a valid comparison. Armstrong and Chandler did not perform detailed calculations on the effects of muon and electron contamination in the experimental beam, but they estimated the effects of
these contaminants and found them to be small, as shown in Fig. 10.
The agreement between the calculated and experimental results in
Fig. 10 is quite satisfactory. Comparisons between calculated and
experimental cell-survival probabilities are a much more stringent
test of the calculations than are comparisons between calculated
and experimental absorbed-dose data because the heavier, highly
ionizing particles contribute very appreciably to the cell-survival
probabilities, while they contribute only a small fraction of the
absorbed doses.

     Additional results on the use of various particles in cancer
radiotherapy will be found in the references given above and in
the work of Alsmiller *et al.*[83]

ACKNOWLEDGMENTS

     Many people have been involved in developing the code HETC
and the capabilities which are used in carrying out calculations
such as those discussed here. Thanks are due particularly to
T. W. Armstrong, J. Barish, H. W. Bertini, K. C. Chandler, T. A.
Gabriel, and R. T. Santoro.

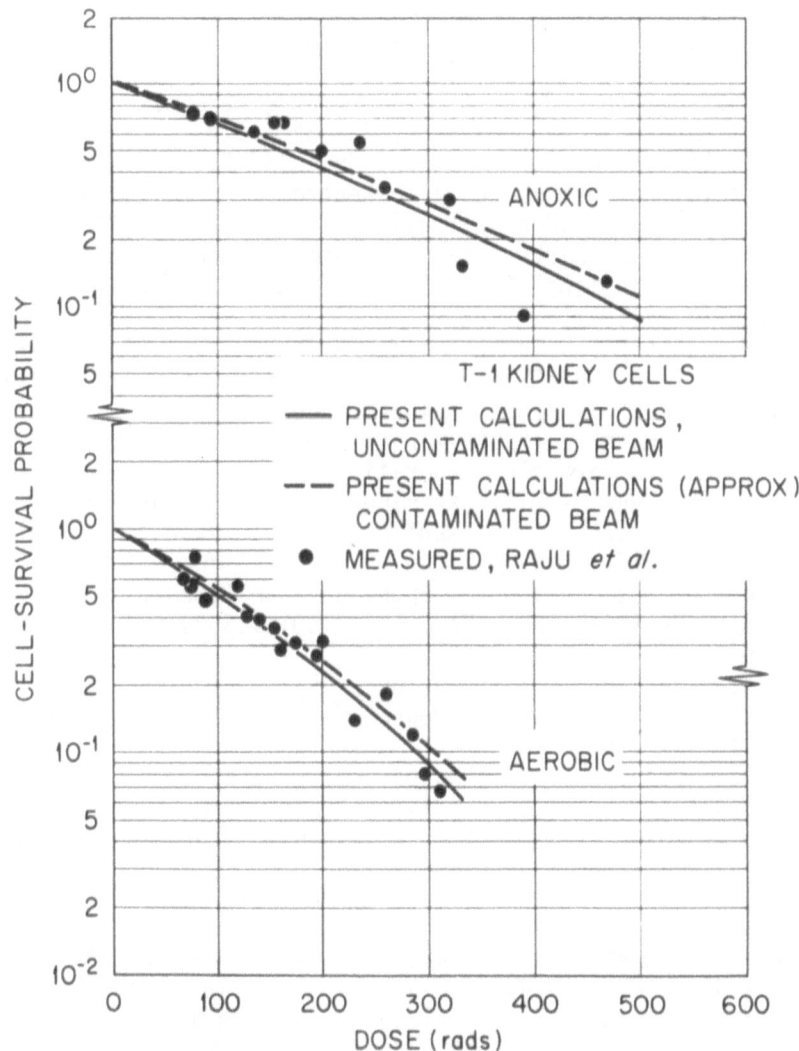

Fig. 10. Comparison of measured[82] and calculated[72] cell-survival probabilities in the capture region of a π^- beam.

REFERENCES

1. K. C. Chandler and T. W. Armstrong, "Operating Instructions for the High-Energy Nucleon-Meson Transport Code HETC," ORNL-4744, Oak Ridge National Laboratory (1972).
2. W. A. Coleman and T. W. Armstrong, "The Nucleon-Meson Transport Code NMTC," ORNL-4606, Oak Ridge National Laboratory (1970).
3. R. G. Alsmiller, Jr., T. W. Armstrong, and W. A. Coleman, Nucl. Sci. Eng. 42, 367 (1970).
4. T. W. Armstrong, R. G. Alsmiller, Jr., K. C. Chandler, and B. L. Bishop, Nucl. Sci. Eng. 49, 82 (1972).
5. D. C. Irving, R. M. Freestone, Jr., and F. B. K. Kam, "05R, a General-Purpose Monte Carlo Neutron Transport Code," ORNL-3622, Oak Ridge National Laboratory (1965).
6. O. Ozer and D. Garber, "ENDF/B Summary Documentation," BNL-17541, National Cross Section Center, Brookhaven National Laboratory (1973).
7. R. W. Roussin, "Defense Nuclear Agency Working Cross Section Library," ORNL-RSIC-34, Vol. 1, Radiation Shielding Information Center, Oak Ridge National Laboratory (1972).
8. Miriam P. Guthrie, "EVAP-4: Another Modification of a Code to Calculate Particle Evaporation from Excited Compound Nuclei," ORNL-TM-3119, Oak Ridge National Laboratory (1970).
9. T. W. Armstrong, K. C. Chandler, and J. Barish, J. Geophys. Res. 78, 2715 (1973).
10. Wilmot N. Hess, Rev. Mod. Phys. 30, 368 (1958).
11. Hugo W. Bertini, Phys. Rev. 131, 1801 (1963) and erratum, Phys. Rev. 138, AB2 (1965).
12. V. S. Barashenkov, Interaction Cross Sections of Elementary Particles, 1966, Israel Program for Scientific Translation, Jerusalem, 1968.
13. T. A. Gabriel, R. T. Santoro, and J. Barish, "A Calculational Method for Predicting Particle Spectra from High-Energy Nucleon and Pion Collisions ($\geq$ 3 GeV) with Protons," ORNL-TM-3615, Oak Ridge National Laboratory (1971).
14. F. D. Becchetti, Jr. and G. W. Greenlees, Phys. Rev. 182, 1190 (1969).
15. C. M. Perey and F. G. Perey, Atomic Data and Nuclear Data Tables 13, 293 (1974).
16. H. W. Bertini, M. P. Guthrie, and O. W. Hermann, "Instructions for the Operation of Codes Associated with MECC-3, a Preliminary Version of an Intranuclear-Cascade Calculation for Nuclear Reactions," ORNL-4564, Oak Ridge National Laboratory (1971).
17. H. W. Bertini, "Monte Carlo Calculations on Intranuclear Cascades," ORNL-3383 (Dissertation), Oak Ridge National Laboratory (1963).
18. V. A. Konshin and E. S. Matusevich, Atomic Energy Review 6, 3 (IAEA 1968).

19. H. W. Bertini, Phys. Rev. 188, 1711 (1969).
20. H. W. Bertini, Phys. Rev. C6, 631 (1972).
21. V. S. Barashenkov et al., Nucl. Phys. A187, 531 (1972).
22. R. G. Alsmiller, Jr. and O. W. Hermann, Nucl. Sci. Eng. 40, 254 (1970).
23. Hugo W. Bertini, Phys. Rev. C5, 2118 (1972).
24. F. E. Bertrand and R. W. Peelle, Phys. Rev. C8, 1045 (1973).
25. H. W. Bertini, G. D. Harp, and F. E. Bertrand, Phys. Rev. C10, 2472 (1974).
26. M. P. Guthrie, R. G. Alsmiller, Jr., and H. W. Bertini, Nucl. Instr. Meth. 66, 29 (1968); with erratum, Nucl. Instr. Meth. 91, 669 (1971).
27. T. A. Gabriel, R. G. Alsmiller, Jr., and M. P. Guthrie, "An Extrapolation Method for Predicting Nucleon and Pion Differential Production Cross Sections from High-Energy (> 3 GeV) Nucleon-Nucleus Collisions," ORNL-4542, Oak Ridge National Laboratory (1970).
28. T. A. Gabriel and R. T. Santoro, "Calculation of the Long-Lived Activity in Soil Produced by 500-GeV Protons," ORNL-TM-3262, Oak Ridge National Laboratory (1970).
29. S. F. Singer, Phys. Rev. Lett. 1, 181 (1958).
30. A. J. Dragt, M. M. Austin, and R. S. White, J. Geophys. Res. 71, 1293 (1966).
31. T. A. Farley and M. Walt, J. Geophys. Res. 76, 8223 (1971).
32. W. N. Hess, H. W. Patterson, R. Wallace, and E. L. Chupp, Phys. Rev. 116, 445 (1959).
33. S. S. Holt, R. B. Mendell, and S. A. Korff, J. Geophys. Res. 71, 5109 (1966).
34. E. Heidbreder, K. Pinkau, C. Reppin, and V. Schönfelder, J. Geophys. Res. 75, 6347 (1970).
35. A. M. Preszler, G. M. Simnett, and R. S. White, Phys. Rev. Lett. 28, 982 (1972).
36. A. M. Preszler, G. M. Simnett, and R. S. White, J. Geophys. Res. 79, 17 (1974).
37. T. A. Gabriel, R. T. Santoro, and R. G. Alsmiller, Jr., Nucl. Sci. Eng. 44, 104 (1971).
38. W. W. Engle, Jr., "A Users Manual for ANISN, a One-Dimensional Discrete Ordinates Transport Code with Anisotropic Scattering," K-1693, Computing Technology Center, Union Carbide Corporation (1967).
39. F. B. McDonald, "IQSY Observations of Low-Energy Galactic and Solar Cosmic Rays," in Annals of the IQSY, Vol. 4, edited by A. C. Strickland, MIT Press, Cambridge, MA, 1969, pp. 187-216.
40. T. W. Armstrong, K. C. Chandler, and J. Barish, "Calculations of Neutron Flux Spectra Induced in the Earth's Atmosphere by Galactic Cosmic Rays," ORNL-TM-3961, Oak Ridge National Laboratory (1972).
41. M. Merker, Phys. Rev. Lett. 29, 1531 (1972).
42. A. Van Ginneken and T. Borak, IEEE Trans. Nucl. Sci., NS-18, 3, 746 (1971).

43. J. Ranft, Nucl. Instr. Meth. <u>48</u>, 261 (1967).
44. T. W. Armstrong and R. G. Alsmiller, Jr., Nucl. Sci. Eng. <u>33</u>, 291 (1968).
45. A. Van Ginneken, Fermi National Accelerator Laboratory, private communication (1971).
46. T. A. Gabriel, J. D. Amburgey, and R. T. Santoro, "Calculated Performance of a Mineral-Oil-Iron Ionization Spectrometer," ORNL-TM-4803, Oak Ridge National Laboratory (1975).
47. W. Selove, University of Pennsylvania, private communication.
48. H. L. Beck, "A Monte Carlo Simulation of the Transport of High Energy Electrons and Photons in Matter," HASL-213, USAEC Health and Safety Laboratory (1969).
49. J. B. Birks, <u>The Theory and Practice of Scintillation Counting</u>, The Macmillan Company, New York, NY, 1964.
50. T. A. Gabriel and K. C. Chandler, Particle Accelerators <u>5</u>, 161 (1973).
51. T. A. Gabriel and J. D. Amburgey, Nucl. Instr. Meth. <u>116</u>, 333 (1974).
52. T. W. Armstrong and R. G. Alsmiller, Jr., Proc. Apollo 12 Lunar Science Conf., Geochim. Cosmochim. Acta, Vol. 2, Suppl. 2, p. 1729, MIT Press, Cambridge, MA (1971).
53. D. Lal and B. Peters, "Cosmic Ray Produced Radioactivity on the Earth," in <u>Encyclopedia of Physics</u>, Vol. XLVI/2, edited by K. Sitte, Springer-Verlag, Berlin, Heidelberg, New York, 1967, pp. 551-612.
54. M. Honda and J. R. Arnold, "Effects of Cosmic Rays on Meteorites," in <u>Encyclopedia of Physics</u>, Vol. XLVI/2, edited by K. Sitte, Springer-Verlag, Berlin, Heidelberg, New York, 1967, pp. 613-631.
55. T. W. Armstrong and R. G. Alsmiller, Jr., Nucl. Sci. Eng. <u>38</u>, 53 (1969).
56. T. A. Gabriel, Nucl. Instr. Meth. <u>91</u>, 67 (1971).
57. T. A. Gabriel and R. T. Santoro, Nucl. Instr. Meth. <u>95</u>, 275 (1971).
58. LSPET (Lunar Sample Preliminary Examination Team), Science <u>165</u>, 1211 (1969).
59. N. Durgaprasad, C. E. Fichtel, and D. E. Guss, J. Geophys. Res. <u>72</u>, 2765 (1967).
60. J. L. Modisette, T. M. Vinson, and A. C. Hardy, "Model Solar Proton Environments for Manned Spacecraft Design," NASA TN D-2746, National Aeronautics and Space Administration (1965).
61. W. R. Webber, "An Evaluation of the Radiation Hazard Due to Solar-Particle Events," D2-90469, The Boeing Company, Seattle, WA (1963).
62. K. C. Hsieh, University of Arizona, private communication (1970).
63. W. R. Yucker, "Statistical Analysis of Solar Cosmic Ray Proton Fluence," MDAC Paper WD 1320, McDonnell Douglas Astronautics (1970).

64. G. Radke, Aerospace Med. 40, 1495 (1969).
65. M. B. Baker, R. E. Santina, and A. J. Masley, A.I.A.A.J. 7, 2105 (1969).
66. ESSA (ESSA Research Laboratories), "Solar-Geophysical Data," Report No. 309, Part II, Environmental Research Laboratories, Boulder, CO (1970).
67. T. W. Armstrong and R. G. Alsmiller, Jr., "Calculation of Cosmogenic Radionuclides in the Moon and Comparison with Apollo Measurements," ORNL-TM-3267, Oak Ridge National Laboratory (1970).
68. J. P. Shedlovsky et al., Proc. Apollo 11 Lunar Science Conf., Geochim. Cosmochim. Acta, Vol. 2, Suppl. 1, p. 1503, Pergamon Press, New York, NY (1970).
69. R. C. Finkel et al., Proc. Apollo 12 Lunar Science Conf., Geochim. Cosmochim. Acta, Vol. 2, Suppl. 2, p. 1773, MIT Press, Cambridge, MA (1971).
70. T. W. Armstrong, J. Geophys. Res. 74, 1361 (1969).
71. T. W. Armstrong and K. C. Chandler, Radiat. Res. 52, 247 (1972).
72. T. W. Armstrong and K. C. Chandler, Radiat. Res. 58, 293 (1974).
73. R. G. Alsmiller, Jr., T. W. Armstrong, and W. A. Coleman, Nucl. Sci. Eng. 42, 367 (1970).
74. R. G. Alsmiller, Jr., T. W. Armstrong, and Barbara L. Bishop, Nucl. Sci. Eng. 43, 257 (1971).
75. T. W. Armstrong, R. G. Alsmiller, Jr., and K. C. Chandler, Phys. Med. Biol. 18, 830 (1973).
76. T. W. Armstrong and B. L. Bishop, Radiat. Res. 47, 581 (1971).
77. R. G. Alsmiller, Jr. and J. Barish, Med. Phys. 1, 51 (1974); with erratum, Med. Phys. 2, 34 (1975).
78. R. Katz, S. C. Sharma, and M. Homayoonfar, Chap. 6 in Topics in Radiation Dosimetry, Suppl. 1 to Radiation Dosimetry, edited by Frank A. Attix, Academic Press, New York, NY, 1972.
79. Robert Katz and S. C. Sharma, Nucl. Instr. Meth. 111, 93 (1973).
80. J. E. Turner et al., Proc. Intl. Congress on Protection Against Accelerator and Space Radiation, April 26-30, 1971, CERN, Geneva, Switzerland, CERN 71-16, Vol. 1, p. 231.
81. J. E. Turner et al., Radiat. Res. 52, 229 (1972).
82. M. R. Raju et al., Brit. J. Radiol. 45, 178 (1972).
83. R. G. Alsmiller, Jr., et al., Radiat. Res. 60, 369 (1974).

NUCLEAR FRAGMENTATION IN THERAPEUTIC AND DIAGNOSTIC STUDIES WITH HEAVY IONS*

A. Chatterjee, C.A. Tobias and J.T. Lyman

Donner Laboratory, Lawrence Berkeley Laboratory,
University of California, Berkeley, California 94720

1. INTRODUCTION

Fast heavy ions previously could be studied only in the outer space where they form important components of primary cosmic rays. Recent developments in accelerator technology have made it possible to obtain energetic heavy ions in the laboratory. In August 1971, the Princeton Particle Accelerator (1,2) and the Berkeley Bevatron (3) produced penetrating deflected beams of nitrogen nuclei. Following these achievements, heavy ion beams of much higher intensity than ever produced before were accelerated in Berkeley in August 1974 by using the existing Super HILAC (HILAC modified to accelerate all ions through uranium to an energy of 8.5 MeV/n) as pre-accelerator and injecting its low-energy heavy-ion beams for synchrotron acceleration in the Bevatron: the BEVELAC project. A complete description of the BEVALAC project is described elsewhere (4). Particles accelerated in BEVALAC can have a maximum energy of 2.1 GeV/nucleon.

Heavy charged particles exhibit excellent depth-dose characteristics, and high LET (linear energy transfer) values after penetrating to depth in tissue. Because of these desirable properties, these beams could be useful in tumor therapy and other applications of deep-lesion production in neurological research and new diagnostic techniques such as laminography and radiography (5). But before such a beam could be applied in medicine with confi-

---

* This work has been supported by the United States Energy Research and Development Administration.

dence, information on beam diagnostics has to be obtained with a great certainty. "Conventional" dosimetry measurements are not sufficient for these applications because they do not give any information on nuclear fragmentation of the high-energy heavy ions as they interact, particularly with tissue. The experience of radiotherapists has been so far limited to low LET radiations (x rays) where absorbed dose is adequate enough for describing a radiation field. With high LET radiations, such as heavy ions, the radiation field consists of a spectrum in energy and particle type, and hence dose is not a sufficient parameter. Because of nuclear fragmentation particles of lower atomic numbers are produced which dilute the main beam. It is quite obvious that a quantitative evaluation of radiobiological data requires information on these fragments.

The phenomenon of nuclear fragmentation is quite well known from the cosmic ray studies. Fragmentation cross sections for the break-up of heavy cosmic-ray nuclei are essential for the interpretation of measured cosmic-ray charge spectra. In general, in a nuclear encounter between a fast nucleus with the resting nuclei of the absorber, one encounters three kinds of phenomena: (i) nuclear stars, the fragments belong to the target and remain more or less localized near the site of interaction; (ii) fast moving secondary nuclei as a result of fragmentation of the incident nucleus; and (iii) the results of head-on collisions between nuclei which are occasionally observed as multipronged nuclear stars. These interactions all depend on the velocity and atomic number of the beam particles involved. Generally more and more secondary particles are produced which accompany the primary beam as it travels deeper into the absorber, the effect maximizing near the region where the incident particle stops. Some secondaries travel on deeper into the absorber, thereby depositing an exit dose. The dose and the biological effect due to these secondaries are expected to be less than the effect from the primary beam particles. In addition, there is a small background of neutrons, gamma rays and, at higher particle velocities, mesons are also produced.

The present report considers the problem of nuclear fragmentation of the incident beam from the point of view of biomedical interest. No attempt has been made to incorporate rigorous physics in quantitative evaluation procedures but, wherever possible, experimental data on fragmentation have been used. As more and more experimental results become available, analysis of radiobiological data will become simpler. From the theoretical point of view, the proton-nucleus interaction is complex enough. Considerable effort was necessary to produce meaningful data on secondary production starting from basic nucleon-nucleon cross sections (6). The problem becomes even more complex when the incident particle is also a nucleus consisting of several bound nucleons. Some, but not

enough, theoretical progress has been made in understanding the
important mechanisms cuasing the fragmentation of such heavy ions
in the energy range of interest (100-1000 MeV/nucleon).

In this report, unless otherwise mentioned, fragmentation
will always refer to fast moving incident nucleus.

2. FRAGMENTATION PARAMETERS IN WATER

From the radiobiological viewpoint we require information on nuclear fragmentation of the projectile using tissue as a target
but, for experimental and theoretical simplicity, tissue can be
approximated by water without introducing much error. Before the
availability of high energy heavy ions in the laboratory, fragmentation parameters were studied by observing cosmic ray interactions
within nuclear emulsion and other targets, but not water (7,8,9,
10,11,12). Maccabee et al. (13) obtained some information on the
fragmentation of 250 MeV/nucleon oxygen-ion beams in water using
silicon semiconductor detectors, but in order to analyze their data
the Maccabee group made certain assumptions which have to be verified through numerous other measurements. We need more such data
on other ions.

In the absence of fragmentation data in water, an effort was
made to extrapolate fragmentation probabilities found in emulsion
through a model first suggested by Noon and Kaplan (14). The
method is general enough to be applicable in any medium.

2.1 Cosmic ray fragmentation data in emulsion

A small amount of data is available in the high energy region from
cosmic ray experiments in nuclear emulsion. The photographic emulsion consists of both heavy and light elements; the average atomic
weight of the heavy elements in emulsion is ∿84 and that of the
light elements (excluding hydrogen) is ∿14. About 1/3 of all
interactions generally occur with light elements.

For evaluating fragmentation probabilities in water, cosmic
ray data (in emulsion) used were those of Fowler, Hillier and
Waddington (15), and Rajopadhye and Waddington (16). In these
compilations the incoming ion was charge-identified and the following information was tabulated on each interaction: the number of
slow prongs; the identity of the fast outgoing heavy fragment
($Z > 2$) in the forward direction (if any); the number of fast heliu
particles (in the forward direction); and the number of "shower"
particles (pions, fast protons, etc.). For nitrogen beam, the
quantitative data are shown in Fig. 1. In this figure no attempt
has been made to show the separate isotopes.

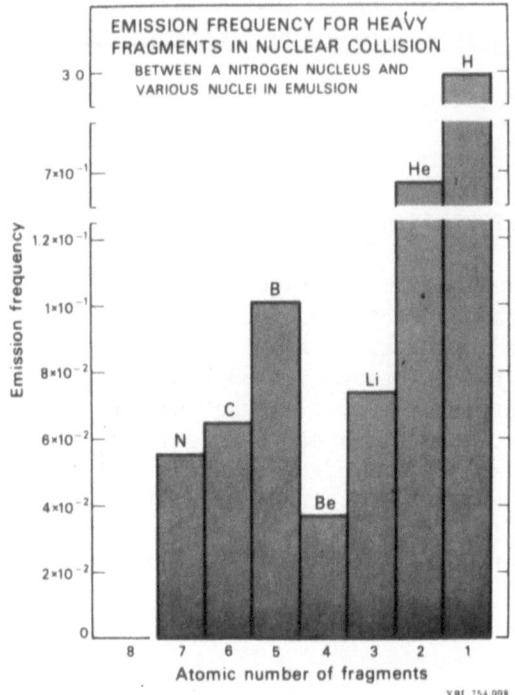

Fig. 1.   Emission frequency of various fast-moving fragments rising
out of a nuclear collision between a fast-moving nitrogen nucleus
and various nuclei in emulsion.   Data obtained are from cosmic ray
studies made by the researchers cited (15,16).   Total nuclear cross
section as well as individual fragmentation cross sections are considered to be independent of energy of the incident particle.

2.2   Extrapolation to water

In order to extrapolate fragmentation data in emulsion to water a
method first adopted by Noon and Kaplan was used.   This method
originates from a rather simple and reasonable idea: in a collision
which involves a large distance (glancing) between an incident
nucleus and a target nucleus, the incoming projectile will not re-
ceive as much excitation as in a more direct collision.   In the
later type of collision one would expect rather complete disintegration giving rise to head-on collision as the extreme limit.
Thus it is the amount of overlap which determines the fragmentation
probability of a particular type of fragment in an interaction.

If we introduce the classical concept of impact parameter as
the distance of closest approach between two interacting nuclei,
then for collisions of all types the impact parameter will vary

between 0 and D, where D is the maximum distance between the centers of the two nuclei at which an interaction can take place. According to Bradt and Peter (17), the total nuclear-interaction cross section, σ_t, is given by the empirical formula

$$\sigma_t = \Pi R_o^2 (A_i^{1/3} + A_t^{1/3} - 2 \cdot \Delta R)^2 \tag{2.2a}$$

where $R_o = 1.45 \times 10^{-13}$ cm, A_i is the mass number of the incident nucleus, and A_t is that of the target nucleus. ΔR is the required overlap before an interaction can take place and is generally assumed to have a value of 0.59×10^{-13} cm. Thus, one can write the maximum impact parameter as

$$D = R_o (A_i^{1/3} + A_t^{1/3} - 2\Delta R) \tag{2.2b}$$

Since all collisions are assumed to occur with impact parameters lying between these limits, the probability of an interaction occurring is

$$P_D = k\int_o^D Xdx$$

$$= 1$$

where k is the normalization constant. The probability of an interaction occurring with an impact parameter between D and D-d is given by

$$P_d = k \int_{D-d}^D Xdx$$

$$\text{i.e., } P_d = [D^2 - (D - d)^2]/D^2 \tag{2.2c}$$

It can easily be seen that for a given incident particle and for a given overlap, d, in two cases: (i) involving a large target nuclei; and (ii) involving a smaller target nuclei; for smaller target nuclei, proportionately more of the effective cross section is utilized than for larger target nuclei. Thus for smaller target nuclei, i.e., smaller D, the proportion of collisions with impact parameters between D and D-d increases. For a mixture of target nuclei, expression (2.2c) can be rewritten as

$$P_d = (\sum_i n_i D_i - \sum_i n_i (D_i - d)^2)/\sum_i n_i D_i^2 \tag{2.2d}$$

where n_i is the number of atoms/cc of the ith type of target nucleus and D_i is the corresponding maximum impact parameter.

For a fixed incoming nucleus (such as nitrogen nucleus) it is reasonable to assume that collisions with impact parameters between D and $D-d_f$ yield fragments of the fth type. In other words, for a fixed incoming nucleus there is a fixed overlap for the production of a given type of fragment. For smaller impact parameters (greater overlap) mainly helium particles and protons are produced.

The overlap parameters d_f, for various fragments can be determined from equation (2.2d) to fit the observed fragmentation probabilities in emulsion. These parameters are then used with a known composition of water to determine the fragmentation probabilities for that medium.

Implicit in the use of the cross-sectional formula given by Bradt and Peter (equation 2.2a) is the approximation that the cross section does not depend on the energy of the heavy nucleus. It is constant despite the slowing down of the projectile. Such an approximation immediately leads to the conclusion that the fragmentation probabilities are also independent of the energy of the heavy nucleus. Hence, under this approximation, the probabilities obtained from the cosmic ray data (high energy heavy nuclei) are also considered to be the same for therapeutic beam energies ($\sim$400 MeV/n) This procedure has been adopted in the absence of an appropriate energy-dependent cross-sectional formula but when this becomes available its inclusion will not affect the method outlined here.

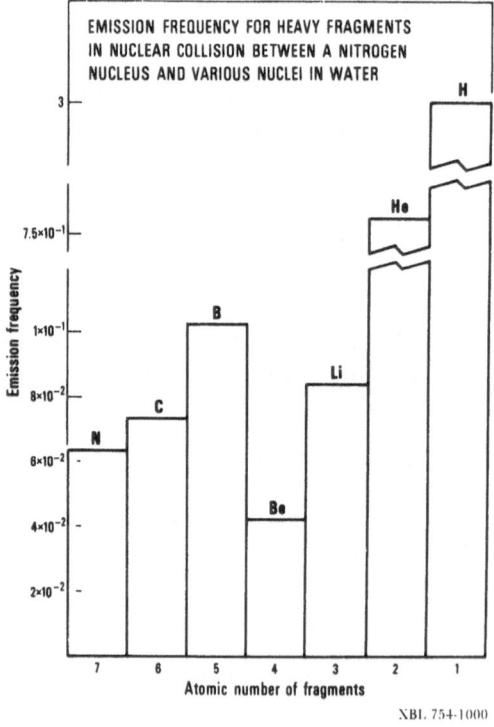

XBL 754-1000

Fig. 2. Cosmic ray data in emulsion have been extrapolated to obtain emission frequency of different fragments produced at high velocities in a collision between a nitrogen nucleus and various other nuclei in water. Fragments discussed in this report are those from the incident particle only.

Following the geometrical model of Noon and Kaplan and using the cosmic ray data in emulsion we have obtained fragmentation probabilities in water for incident nitrogen and oxygen nuclei. These are shown in Fig. 2 and Fig. 3. A comparison between the fragmentation probabilities for nitrogen nucleus in emulsion (see Fig. 1) with that in water for the same nucleus (see Fig. 2) shows that all the respective probabilities are greater in the case of water. This is reasonable in view of the fact that for smaller target nuclei more of the effective cross section is utilized than for larger target nuclei (emulsion).

2.3 Evaluation of fragmentation parameters (Bragg ionization curve)

In the absence of detailed knowledge of experimental data on heavy nuclei fragmentation, an indirect method of evaluating these estimated results is the comparison of current dosimetric measurements of the Bragg ionization curve with predictions based on our extrapolation of fragmentation data.

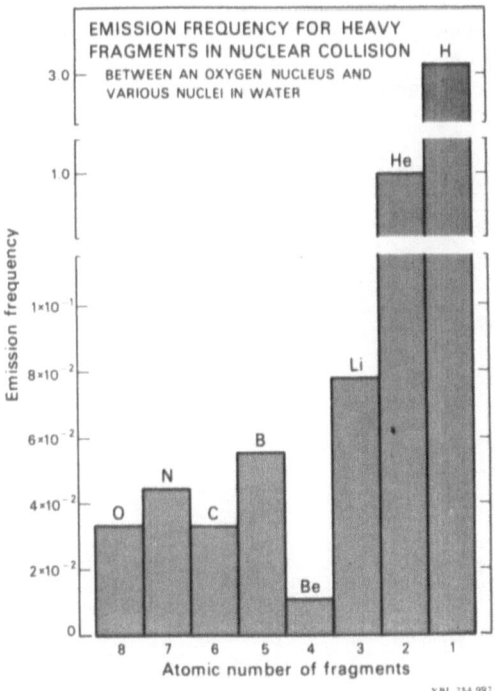

Fig. 3. Emission frequency for different fragments produced as a result of breakup of oxygen nucleus in nuclear interaction with various nuclei of water. This is a result of extrapolation from the fragmentation data in emulsion.

The dose calculations were made for 260 MeV/n nitrogen and also for 230 MeV/n oxygen ions incident on a water absorber. A number of assumptions were introduced for ease of calculation. They are as follows:

1) The energetic heavy fragment (if any) leaving the interaction site was assumed to be the residual part of the incident particle and was considered to be moving in the forward direction with the same velocity as the incident particle. This assumption may not be valid at lower energies. This is a very standard assumption and has been considered by many in experimental data analysis (18). Sometimes this is known as "straight-ahead" approximation.

2) The fast helium particles were also assumed to be traveling with no change in velocity or direction.

3) The interaction cross section was assumed constant as a function of energy.

4) Tertiary interactions were neglected.

For biomedical interest, two of the most important dosimetric considerations are the number-distance relation (differential range curve) and the nature of the depth-dose distribution. With present fragmentation parameters and the above mentioned assumptions, the computer code BRSEC (19) was used to calculate the number-distance curve for nitrogen as well as oxygen beams. A typical result for

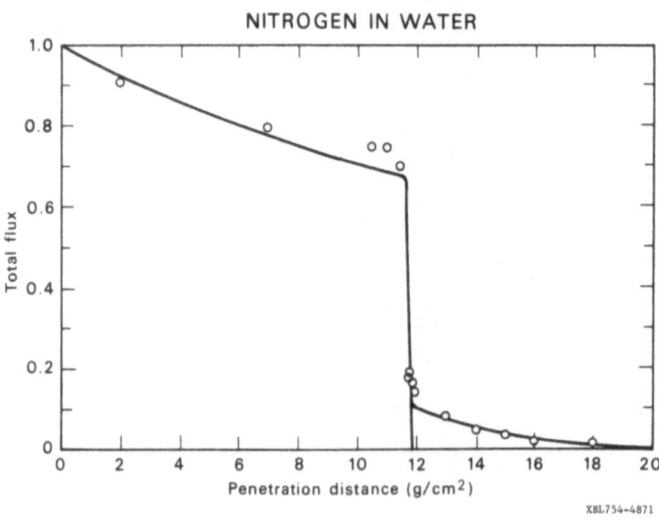

Fig. 4. Number/distance curve for a 260 MeV/n nitrogen nucleus penetrating through water. The solid-line curve is estimated from the present fragmentation data in water. Experimentally obtained point are represented by circles.

nitrogen beam is shown in Fig. 4. The theoretical curve is shown
by the solid line and experimentally measured points (circles) are
also indicated. Multiplicating the fluence of each species at
various depths by the corresponding LET values yields a value for
density of energy transferred (i.e., absorbed dose) so that a
depth-dose relation can be calculated. Using such a procedure a
calculation was made for both oxygen and nitrogen beams (Figs. 5
and 6). As before, experimentally measured values are also shown
on the same plots. Agreement between calculation and experiment
is rather good in view of the various assumptions made in the calculations. It seems that for radiobiological purposes the method
of obtaining fragmentation parameters by extrapolating from one
medium to another is quite adequate, at least with respect to Bragg
ionization phenomenon.

3. DOME DISTRIBUTION

In cancer oriented pretherapeutic studies sometimes it is very im--
portant to obtain information on the nature of fluence distribution
of different particles at various depths in the absorber. We have
made some theoretical estimates of the distribution of forward

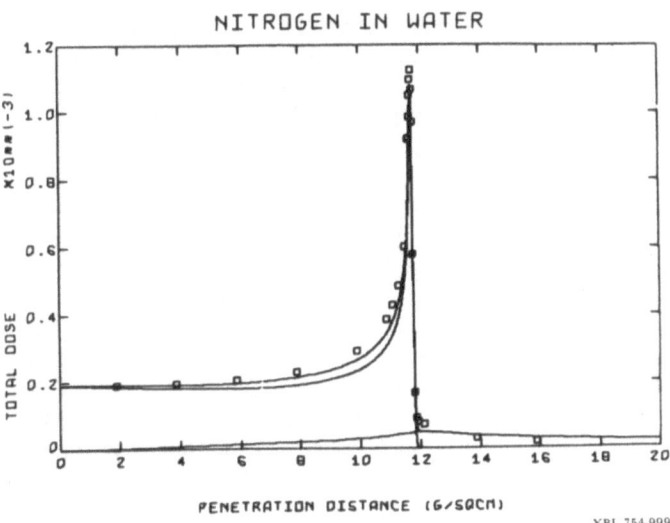

XBL 754-999

Fig. 5. A typical Bragg ionization curve estimated with respect to
extrapolated fragmentation data (see Fig. 2) for a 260 MeV/n nitrogen nucleus in water (solid-line curve). The squares represent experimentally measured values. Primary and secondary contributions
to total ionization are also shown.

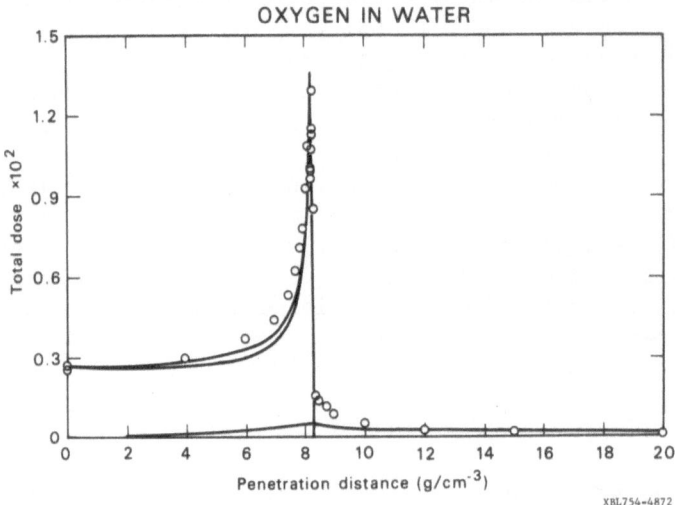

Fig. 6. Estimated Bragg ionization curve for 230 MeV/n oxygen in water. Secondary contribution to dose due to various fragments has been obtained from the present extrapolated data. Primary contribution comes from the unfragmented incident nucleus. Experimentally measured values are indicated by circles.

secondaries in monoenergetic heavy ion beams. Because of their characteristic shape, we call these "dome distributions".

Let $\phi_p(x)$ denote the fluence of primary heavy ions at a penetration depth x. Through nuclear interaction some of the incident particles are removed from the primary beam. Let the macroscopic total cross section for removal of a primary from the incident beam be represented by σ_p. For a medium of mixed nuclear species, e.g., water

$$\sigma_p^{H_2O} = N_H \sigma_p^H + N_O \sigma_p^O \tag{3a}$$

where N_H is the number of hydrogen atoms present per unit volume and σ_p^H is the macroscopic total cross section for primary beam interaction with hydrogen. Similarly, N_O is the number of oxygen atoms per unit volume and σ_p^O is the primary beam interaction cross section. The rate of removal of primary particles from the beam is given by

$$d\phi_p/dx = \sigma_p \phi_p (x) \tag{3b}$$

A solution of this equation

$$\phi_p (x) = \phi_p (o) \exp (-\sigma_p x) \tag{3c}$$

represents the primary fluence at a depth x. Due to the loss of primary particles, secondary fragments are produced and the rate of production of type f fragments is given by

$$d\phi_f (x)/dx = \phi_p \sigma_{pf} \tag{3d}$$

where $\phi_f (x)$ is the secondary fluence at a depth x and σ_{pf} denotes the macroscopic partial cross section for production of secondary fragments of type f.

In the event that the loss of secondary fragments through nuclear interaction can be neglected, the solution of equation (3d) can be given as

$$\phi_f (x) = \phi_p (o) [1 - \exp(-\sigma_p x)] [\sigma_{pf}/\sigma_p] \tag{3e}$$

The ratio σ_{pf}/σ_p represents the frequency that a fragment of type f is produced per collision, which removes a primary.

Since both the primary particles and the secondary fragments are charged particles, they interact very strongly with the medium electrons and hence loose energy to the medium. Thus, they have a definite range in a stopping medium depending on their charge and kinetic energy. If we neglect range straggling, which is not far from reality, primaries which survive nuclear interactions are finally removed at range R_p, i.e.,

$$\phi_p (x) = 0 \quad \text{for } x > R_p \tag{3f}$$

Secondary fragments are assumed to be produced at the same velocity the primary particles have at the point of fragment production. These fragments can have greater or smaller ranges than the residual range of the primary beam. Generally they will have a greater range than the primary, since ranges of fast heavy charged particles (at the same velocity) scale as (A/Z^2) where A is the atomic mass and Z is the nuclear charge. But, in the case of a neutron stripping interaction, the charge remains the same so the range of the secondary is less than that of the primary from the point of production.

If we denote the ratio of the range of the secondary fragment to the range of the primary as η_f we have from the simple scaling law

$$\eta_f = R_f/R_p = (A_f/A_p)(Z_p/Z_f)^2 \tag{3g}$$

Considerations such as these lead us to the conclusion that there is a build-up in the secondary fluence by fragmentation of primaries until range R_p, where all the primaries disappear. Following this, there is a continuous drop of the secondary fluence as they are removed at the end of their respective ranges. The maximum penetration of a type f fragment will be given by its own range. Fragments produced deeper into the medium will have smaller penetrations, the minimum penetration being that of a fragment produced just as the primary energy is less than the coulomb barrier. Since a majority of the fragments will penetrate deeper than the primary, it is of interest to calculate the flux of such secondaries beyond the primary range.

But the fragments of type f are produced at an arbitrary penetration depth x; the range of these fragments will be given by $\eta_f(R-x)$, where (R-x) is the residual range of the primary particles at the point of their interaction. If we assume that the fragments are not lost through nuclear interaction (tertiary process), then they will stop at a depth x', given by

$$x' = x + \eta_f(R-x) \tag{3h}$$

Through a procedure similar to the one outlined before, the secondary fluence at any depth beyond R, the range of the primary beam, it can be shown that

$$\phi_f(x) = \phi_p(o)[1-\exp(-\sigma_p(\eta_f R-x)/(\eta_f-1))][\sigma_{pf}/\sigma_p] \tag{3i}$$
$$\text{for } x \geq R$$

Equation (3e), which is valid for $x \leq R$, and equation (3i), which is valid for $x \geq R$, represent the theoretical analysis of the dome distribution. There will be several different types of fragments generated producing a family of dome distribution curves, characterized by different values of the maximum range $\eta_f R$, and the maximum fluence, proportional to the parameter (σ_{pf}/σ_p).

From the flux distribution of secondaries one can easily calculate the dose distributions at various depths by multiplication of the fluence by the LET corresponding to their residual energy at any given depth. An example of such a calculation is shown in Fig. 7, which demonstrates the dome distribution for various secondary fragments generated by a 275 MeV/n nitrogen-14 ion beam penetrating in water. The range of such a beam is 12.55 cm. Dome distributions were computed using the fragmentation parameters (σ_{pf}/σ_p) and range ratios shown in Table I. The fragmentation parameters used are fictitious, but the general correctness of the distribution would not be altered if correct parameters were available. One cannot obtain specific data on various isotopes from the emulsion data. The fluence of each secondary component is plotted

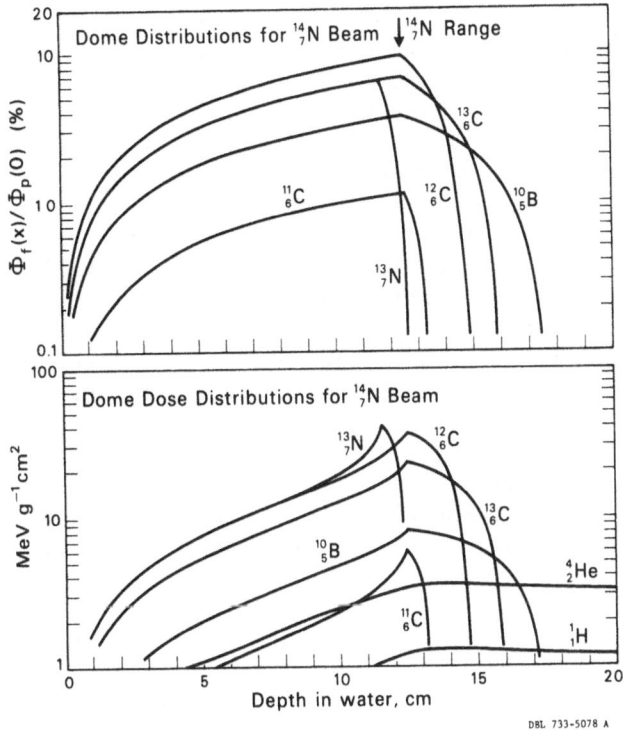

Fig. 7. <u>Top curves</u>: Estimated values for fluence distribution, showing fluence of nuclear secondaries generated by a $^{14}_{7}N$ beam of 12.5 cm range. Heavy fragments come to a stop near the range of the parent beam. <u>Bottom curves</u>: Calculated values for "dome-dose" distribution due to secondary fragments in the $^{14}_{7}N$ beam. Various peaks appear in close proximity to the Bragg peak of the primary beam.

Table I.

Fragment	$^{13}_{7}N$	$^{13}_{6}C$	$^{12}_{6}C$	$^{11}_{6}C$	$^{10}_{5}B$	$^{4}_{2}He$	$^{1}_{1}H$
σ_{pf}/σ_{p} (not normalized)	0.18	0.18	0.25	0.03	0.10	0.50	0.75
Range Ratio (η_{f})	0.93	1.26	1.17	1.07	1.40	3.50	3.50

on the upper graph of Fig. 7 as a fraction (percent) of the incident primary fluence, against depth of penetration in water. The
arrow indicates the range of the incident primaries. The lower
curve in Fig. 7 represents the composite dose distributions of the
secondary nuclear component as a function of depth in water. Because it contains stopping particles of nitrogen, carbon and boron
near the Bragg peak of the primary beam, the average LET is fairly
high there. Dose contributions from head-on collisions are small
because of the small size of cross sections of the particles, and
dose contributions from local stars are also relatively small since
these are distributed along the entire beam range.

An example of the application of the dome distribution to real
data is shown in Fig. 8. The closed circles are experimental data
measured by Maccabee et al. (13,18). The ordinate is the number
of carbon secondaries identified at a given depth of penetration
in water, normalized per 26,769 oxygen ions incident at 250 MeV/nu-
cleon. The solid line is a theoretical fit to the data using the
dome distribution expression. $\sigma_{o \rightarrow c}$ was varied for best fit,

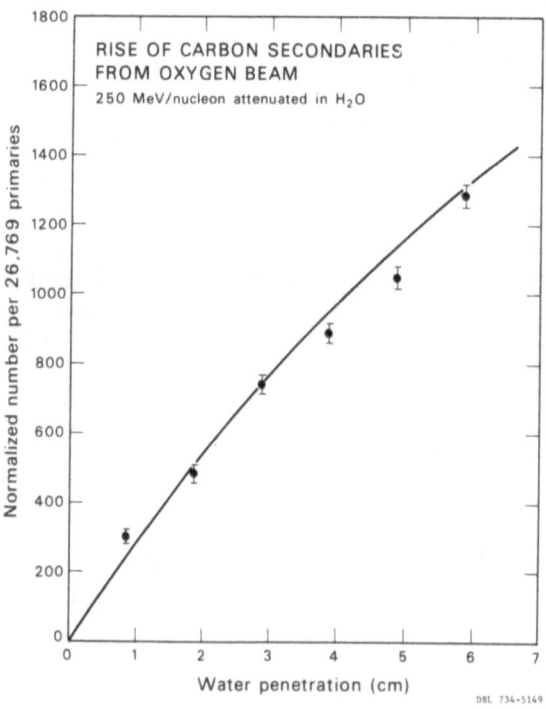

Fig. 8. One section of the dome distribution curve representing
the rise of carbon secondaries before the range of the parent beam,
250 MeV/n oxygen ions in water. Solid circles represent the experimentally measured values obtained by Maccabee et al. (13).

yielding a value of the microscopic partial cross section for a water molecule

$$\sigma_{o \to c}^{H_2O} = 324 \pm 64 \quad mb \tag{3j}$$

No experimental results are currently available to verify the dome distribution expression beyond the range of the incident particle.

4. AUTOACTIVATION

Fragments which are radioactive (the phenomenon of autoactivation) will have useful applications in the field of nuclear medicine and in diagnostic procedures related to radiation therapy. It has been demonstrated that a fraction of heavy fast particle beams become radioactive as they pass through matter. These radioactive fragments are energetic enough to penetrate a reasonable depth in tissue for nuclear medicine and therapy. We believe that the recently discovered phenomenon of "autoactivation" (20) is quite general for fast heavy nuclei and also occurs in heavy primary cosmic radiation.

We shall illustrate autoactivation by considering a high energy oxygen beam. As indicated in Fig. 9, an individual oxygen particle will undergo collisions with resting nuclei in the absorber. It has been mentioned earlier that, in general, the fast moving particle often becomes fragmented, with each of the fragments continuing to move without appreciable change of direction or velocity except for a few MeV in energy exchange. After slowing down, these fragments eventually deposit along the tracks of oxygen nuclei. If a specific fragment is a radioactive isotope, we call the phenomenon "autoactivation" or sometimes autoradioactivation. These radioactive particles will decay for some time (depending upon their half-life) after stopping within the absorber. If they are produced with sufficient cross section, they can be detected quantitatively by proper counting arrangements.

In order to demonstrate such a phenomenon we used a nearly parallel beam of 2.7×10^6 nitrogen particles of energy 270 MeV/n from the Berkeley Bevatron. We allowed the particles to be stopped by a slab of spectroscopically pure beryllium metal (impurities less than 0.1%). Since we were interested in those fragments which decay by positron emission, gamma ray spectra from the irradiated sample were studied with a scintillation spectrometer. The scintillation spectrometer consisted of an 8-inch diameter by 4-inch thick NaI(Tl) crystal and a digital gain stabilizer 400-channel pulse height analyzer. The very low and constant background of the spectrometer system (48 counts per minute in the 511 keV peak region) enabled precise measurement of the small counting rate encountered. A detailed description of the spectrometer is given elsewhere (21).

NUCLEAR FRAGMENTATION

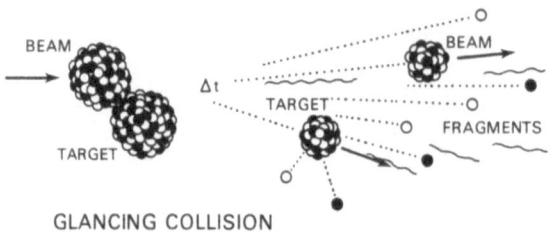

GLANCING COLLISION

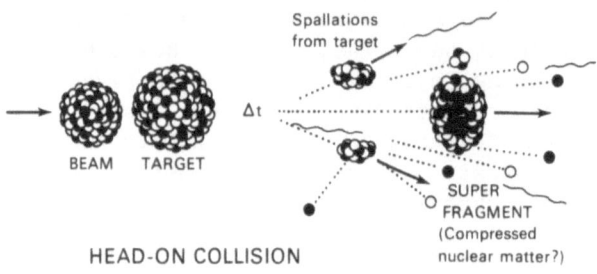

HEAD-ON COLLISION

XBL 7411-8134

Fig. 9. Schematic representation of the collision process of a fast moving oxygen-16 particle with a resting oxygen-16 nucleus in the target. Radioactivity can be detected from two separate regions: (a) at the site of the collision, producing target activity; and (b) at the region where the parent beam comes to stop, giving rise to "autoradioactivity".

No appreciable radioactivity was detected in the irradiated slab of spectroscopically pure beryllium metal, except in the channels set to measure annihilation radiation at 0.51 ± 0.01 MeV. The data obtained in these channels were analyzed into two components with half-lives of 20.34 minutes and 10.1 minutes, identifying the production of carbon-11 and nitrogen-13 respectively, as a result of the fragmentation of the incident beam only. The data for carbon-11 are plotted in Fig. 10 as a typical example. The cross section for production of carbon-11 and nitrogen-13 from incident nitrogen-14 beams directed at the beryllium target was obtained by analysis of the activity data and incident flux. We find the cross section for $^{14}_{7}N \rightarrow ^{11}_{6}C$ is 17 ± 4 millibarns and that for $^{14}_{7}N \rightarrow ^{13}_{7}N$ it is 6 ± 1.5 millibarns. Because of some amount of uncertainty in the measurement of the flux of the initial nitrogen beam, the absolute cross sections may not be very accurate compared to the

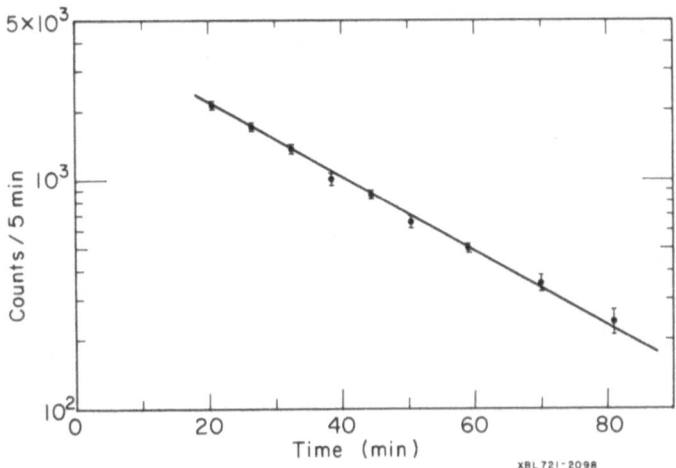

Fig. 10. A semilog plot of activity (counts/5 min) against time (min) is shown. The slope of the line corresponds to ^{11}C activity resulting when 250 MeV/n nitrogen ions are stopped in a beryllium target.

relative cross sections. It is interesting to note that when Heckman et al. (22) used hydrogen as a target, they obtained cross sections for production of carbon-11 and nitrogen-13 as 10.4 milli-barns and 3.6 millibarns respectively for an incident nitrogen-14 beam. Qualitatively, there is an agreement between our results and those obtained by Heckman et al. Furthermore, it is reasonable to expect that the cross sections would be somewhat smaller using hy-drogen as a target compared to beryllium. Another interesting fea-ture is that no matter what target is used, the cross section ratio between carbon-11 production and nitrogen-13 production is the same. For example, whether the target is hydrogen or beryllium, the ratio is 2.8.

In another experiment, the depth-activation distribution was measured for carbon-11 activity by directing a collimated beam of nitrogen particles towards a stack of beryllium absorbers each 0.5 cm thick. Measurement of activity in each disc enabled us to determine the amount of carbon-11 deposition at various depths. The results of this experiment are shown in Fig. 11. No measurable radioactivity could be detected in the first 10 gm/cm^2 of beryllium. The radioactivity in the 0.51 meV channel was at background level

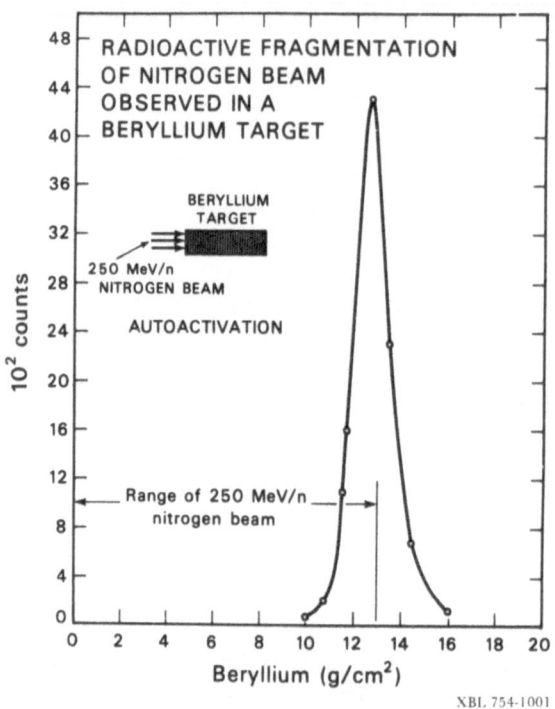

Fig. 11. Plot showing distribution of ^{11}C activity at various depths in a beryllium target for an incident nitrogen beam of range 13 g/cm^2 in beryllium.

or smaller. Most of the carbon-11 activity was concentrated in a fairly sharp peak near the range of nitrogen-14 particles. The range of 250 MeV/n nitrogen ion in beryllium has a range of 13.0 gm/cm^2. Again, no activity was found beyond 16.0 gm/cm^2.

In a similar experiment, a 220 MeV/n oxygen-16 beam was allowed to impinge on a Lucite (tissue-like material) target. Activity of oxygen-15 (half-life 2.1 minutes) produced as a result of beam fragmentation was measured at various depths. The results are shown in Fig. 12. The location of the peak of radioactivity can be compared with the Bragg peak ionization curve of the oxygen nuclei (dashed line). It is indicative that, by measuring the shape and location of such radioactivity distributions through the use of external gamma ray counters and coincidence techniques, it is possible to determine the range penetration of the parent oxygen-16 nuclei from which the autoradioactive species was formed. Note that the "target" radioactivity of ^{15}O produced in the resting nuclei of the absorber (Lucite contains oxygen-16) is relatively small compared to "autoactivity".

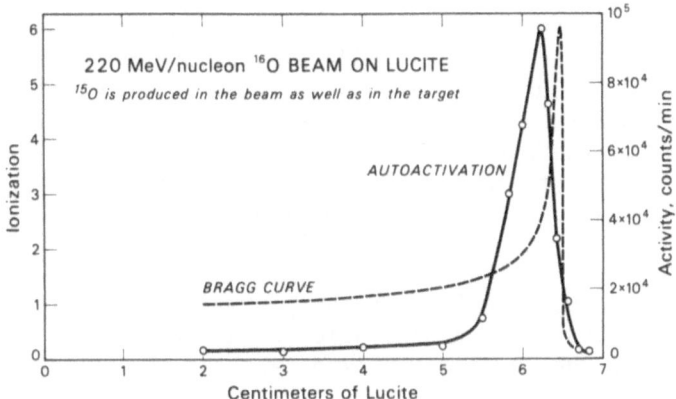

Fig. 12. The distribution of radioactive fragments (^{15}O) at various depths in a Lucite absorber for an incident ^{16}O beam (220 MeV/n) is shown by the continuous line. In addition to the autoactivation peak, a constant activity is seen due to target activation. The Bragg ionization curve is represented by the dashed line.

We understand the general features of the physics associated with the phenomenon of autoactivity. The peak in autoactivity corresponds to the fragmentation of the main beam just at the point of entrance in the target. The range penetration (R_f) of the fragments depends on the quantity Z_f^2/A_f for each fragment. Here, Z_f is the atomic number and A_f is the isotopic mass number for the fragment. If fragment and parent oxygen atoms have the same velocity at the moment when the fragment is produced, then

$$\frac{R_f}{R_o} = \frac{Z_o^2 \cdot A_f}{Z_f^2 \cdot A_o} \tag{4a}$$

where the subscript o refers to incident particle. When an oxygen-15 fragment is produced from an oxygen-16 nucleus, the range of this fragment produced just at the entrance of the target is

$$R_f = \frac{15}{16} R_o \tag{4b}$$

Therefore the location of the autoactivation peak and the Bragg
ionization peak should be related.   This is verified in Fig. 12.

This relation is not strictly correct for all fragments.   From
our experimental measurements we find that deviation from this rule
occurs when the change in nucleon number is more than one.   In
other words, if the nucleon numbers of the fragments differ from
that of the parent nucleus by more than one, the location of the
autoactivity peak occurs at a smaller depth than given by equation
(4a).   Fig. 13 shows the autoactivation peaks of different fragments.   The location of the autoactivation peak will be useful in
diagnostic procedures (see Discussion section) related to radiation
therapy.

Since collisions occur all along the path of the oxygen-16
nuclei, we actually observe a distribution of activity.   The atomic
number of the target nucleus is only important for the yield of a
given fragment.

Since the magnetic rigidity of the fragments produced is usually different from that of the parent beam, it is possible to generate radioactive fragments of therapeutically adequate energy in
thin absorbers and to magnetically separate the fragment beam from

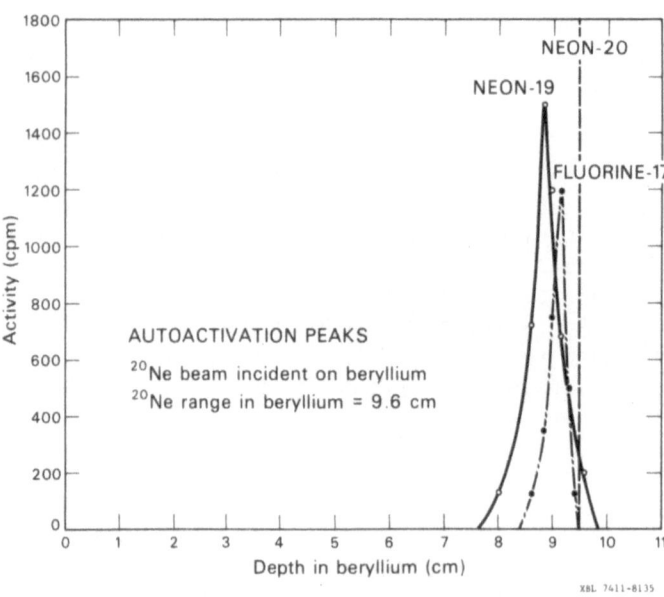

Fig. 13.   Autoactivation peaks for various fragments as identified
through decay curve are shown for an incident neon-20 nucleus in
beryllium.   Range of the beam in beryllium was 9.6 cm.

the main one, as well as other fragments. Working with the physics group of Heckman and Grunder, a small deflected beam of oxygen-15 (approximately 80% purity) was produced at the Bevatron in September 1972.

More experiments on the phenomenon of autoactivation are in progress at the BEVALAC. Even though the fragmentation cross sections are small, the BEVALAC produces very high intensity beams. The availability of a diagnostically meaningful intense beam of radioactive particles of high energy is therefore a distinct possibility in future.

5. DISCUSSION

From the standpoint of radiobiology, radiation therapy and diagnostic procedures, a quantitative knowledge of nuclear fragmentation is essential. Radiobiological end points are strongly dependent of the charge, mass, flux and velocity of a heavy particle beam. The effect of nuclear fragmentation is to dilute the main beam with lower charge (and/or mass) number particles. Thus, in any irradiation arrangement with heavy nuclei, there will be a heterogeneous field of radiation. Fragmentation studies should provide information on such a mixed field of radiation.

In the past, fragmentation parameters have been studied by observing cosmic ray interactions with nuclear emulsion. Such a technique suffers both from limited numbers of interactions and the difficulty of precisely determining the charge and energy of the incident particle. These difficulties are completely avoided in dealing with accelerator produced heavy ions. It is expected that more accurate quantitative information on nuclear fragmentations will be forthcoming from numerous experiments to be performed at the BEVALAC.

So far the bulk of information available on fragmentation is at cosmic ray energies. For radiation therapy and radiological diagnostic considerations, experimental measurements as well as theoretical analysis must be done at relatively lower energies (<500 MeV/n) of incident heavy nuclei. At these low energies the interaction cross section is expected to depend on the energy of the particle. There is not enough information on this dependence presently available. Furthermore, the "straight-ahead" approximation which is generally made for simplicity may introduce large error at low energies for its breakdown.

It is not clear which heavy nucleus is ideal for therapy. Some believe that a heavy nucleus such as neon (Z = 10), or even a heavier nucleus such as argon (Z = 18), may be adequate, but a complete evaluation will not be possible unless information on their

fragments is available. The data available at present from the
cosmic ray studies are inadequate statistically. From the fragmentation data presented in this report on nitrogen and oxygen,
secondary contribution to the dose is small; however this can
hardly be expected for neon nucleus and definitely not for argon
ions.

Heavy ions are presently being considered for localizing the
co-ordinates of a tumor. In our laboratory a technique is under
study which we call "heavy-ion laminography" since it can furnish
density distribution of tissues. In September 1972 the first experiment in heavy-ion radiography using Bevatron-accelerated oxygen
ions was carried out by Benton and Tobias (23).

Because of the steep depth-dose relationships of heavy-ion
beams, it is important to be able to define the co-ordinates and
depth of penetration at each treatment position. X-ray and radioisotope diagnostic methods usually supply inadequate depth information for this purpose. Heavy ions undergo negligible straggling
and scatter, considerably less than even protons. Hence, for a
given differential in density resolution, a lesser number of heavy
particles is needed. But, due to nuclear fragmentation, the required flux of these heavy particles has to be increased somewhat.
Initial considerations of straggling, scattering and fragmentation
make a beam of carbon nucleus quite suitable for radiography or
laminography. The diagnostic dose for such a beam is kept at the
minimum possible. There is no special advantage of using much
heavier nuclei for these purposes.

The phenomenon of autoactivation will be very useful in localizing Bragg peak in patients. The lack of exact diagnostic knowledge on the position of Bragg peak has hindered therapeutic procedures. Exact calculation has been impractical because of intervening bone, tissues, sinews and, in some locations, air. Autoactivity of accelerated heavy particles produces a localized peak
of radioactivity in the region where the primary particles stop,
in an exact measurable relationship to the location of Bragg ionization peak. In other words, it should be possible to apply a suitable coincidence counting technique to measure the range-energy
relationship for a radioactive beam. If the nucleon number of such
radioactive particles does not differ from that of the parent nucleus therapeutic beam by more than one, the range-energy relation
for the beam can be easily derived from the scaling law given by
equation (4a).

Since the charge to mass ratio of a desired radioactive fragment is different from the rest, it can be separated from the main
beam by deflection in appropriate magnetic channels. With the operation of BEVALAC it is now possible to obtain a relatively intense
and diagnostically adequate "carrier-free" radioactive beam to be

used in radiobiological studies and therapy. Besides applications in therapy we anticipate that the use of radioactive beams will find important diagnostic applications in nuclear medicine.

REFERENCES

1. White, M.G., Isaila, M., Prelec, K. and Allen, H.L. Science 174, 1121 (1971).
2. Isaila, M.V., Schimmerling, W., Vosburgh, K.G., et al. Science 177, 424 (1972).
3. Grunder, H.A., Harsough, W.D. and Lofgren, E.D. Science 174, 1128 (1971).
4. Grunder, H.A. Heavy Ion Facilities and Heavy Ion Research at Lawrence Berkeley Laboratory, Lawrence Berkeley Laboratory Report No. 2090 (1973).
5. Benton, E.V., Henke, R.P. and Tobias, C.A. Science 182, 474 (1973).
6. Bertini, H.W. Oak Ridge National Laboratory Report No. ORNL-TM-1996 (December 1967).
7. Rajopadhye, V.Y. and Waddington, C.J. Phil. Mag. 3, 19 (1957).
8. Judek, B. and van Heerden, I.J. Canad. J. Physics 44, 1121 (1966).
9. Waddington, C.J. Progress in Nuclear Physics, ed. by O.R. Frisch, Pergamon Press, New York (1960). 8, 1-46.
10. O'Dell, F.W., Shapiro, M.M. and Stiller, B. J. Phys. Soc. (Japan Suppl. A-111), 17, 23 (1962).
11. Friedlander, M.W., Neelakantan, K.A., Tokunaga, S., Stevenson, G.R. and Waddington, C.J. Phil. Mag. 8, 1691 (1963).
12. Badhwar, G.D., Durgaprasad, N. and Vijayalakshmi, B. Proc. Ind. Acad. Sci. 61, 374 (1965).
13. Maccabee, H.D. and Ritters, M.A. Rad. Res. 60, 409 (1974).
14. Noon, J.H. and Kaplan, M.F. Phys. Review 97, 769 (1955).
15. Fowler, P.H., Hiller, R.R. and Waddington, C.J. Phil. Mag. 2, 293 (1957).
16. Rajopadhye, V.Y. and Waddington, C.J. Phil. Mag. 3, 19 (1958).
17. Bradt, H.L. and Peters, B. Phys. Rev. 77, 54 (1950).
18. Maccabee, H.D. Rad. Res. 54, 495 (1973).
19. Litton, G.M. BERSEC is a Modification of Code BRAGG, Lawrence Radiation Laboratory Report No. UCRL-17392.
20. Tobias, C.A., Chatterjee, A. and Smith, A.R. Physics Letters 37A, 119 (1971).
21. Lawrence Berkeley Laboratory Health Physics Staff. Accelerator Health Physics Training Manual, Lawrence Berkeley Laboratory Report No. 18222 (1968).
22. Heckman, H.H., Greiner, D. and Lindstrom, P. Science 174, 1130 (1971).
23. Tobias, C.A. and Benton, E.V. Report UCID-3593, Berkeley, California (1972).

COSMIC RAY PRODUCED NEUTRONS AND NUCLIDES IN THE EARTH'S ATMOSPHERE*

R. E. Lingenfelter

Department of Astronomy, Department of Geophysics
and Space Physics, Institute of Geophysics and
Planetary Physics, University of California, Los
Angeles, California 90024

Nuclear interactions of cosmic rays in the earth's atmosphere produce an enormous variety of secondary particles, nuclides and radiation: neutrons, protons, antinucleons, electrons, positrons, neutrinos, mesons, hyperons, stable nuclides, radionuclides, X-rays and gamma rays. Although the production of these various secondaries are thus intimately related, the general problem has never been attacked as a whole, nor do I find that it has ever been reviewed as such. There are, however, a number of relatively recent reviews covering various aspects of the problem, especially those relating to neutrons and radionuclides. I shall therefore discuss only in passing studies already covered in previous reviews and concentrate on three specific areas. Two are areas in which significant new contributions have recently been made: in the understanding of the role of cosmic ray produced neutrons as a source of radiation belt particles and in the study of geophysical and astrophysical variations in the cosmic ray production rate of C^{14}. The third area is one which has not yet been adequately explored, but is one which I think holds great promise in its astrophysical, geophysical and even biological implications. That is the study of cosmic ray produced Be^{10} and Al^{26} activity in deep sea sediments, which can provide a record of the global averaged cosmic ray density for the last ten million years - giving important new information on variations of the geomagnetic field, on the long term solar activity, on the origin and propagation

---

*This work was supported in part by the National Science Foundation.

of cosmic rays and on the past variations in the biological
radiation dosage.

Before reviewing these specific areas, however, I will
briefly mention the previous review papers covering different
aspects of the general problem of cosmic ray interactions in the
atmosphere. The measurements and calculations of the cosmic ray
neutron distribution in the atmosphere have been extensively
treated by Haymes (1965) and Schopper et al. (1967). Recent
studies of the cosmic ray neutron leakage current above the
atmosphere and its role as a source of radiation belt particles
have been reviewed by Lockwood (1973) and White (1973), respectively. The general study of cosmic ray produced radioisotopes
in the atmosphere has been reviewed by Lal and Peters (1967)
and Oeschger et al. (1971), while production of radiocarbon with
particular attention to time variations was treated by Lingenfelter and Ramaty (1971). A recent review of interest to the
general problem of cosmic ray interactions in the atmosphere is
that of Daniel and Stephens (1974) treating cosmic ray produced
electrons and gamma rays in the atmosphere.

Experimental and theoretical interest in cosmic ray produced neutrons and radionuclides in the atmosphere has been
primarily stimulated by the importance of radiocarbon dating
(Libby, 1946) to such a wide range of fields and by the desire
to understand the role of cosmic ray neutrons as a source of
particles in the radiation belts (Singer, 1958; Hess, 1959; and
Kellogg, 1959). Since natural radiocarbon is produced mainly by
the capture of cosmic ray neutrons on atmospheric nitrogen
$N^{14}(n,p)C^{14}$, both interests led to extensive measurements and
calculations of the distribution of cosmic ray neutrons in the
atmosphere.

The measurements of the energy spectrum and distribution of
cosmic ray produced neutrons in and above the atmosphere, starting with the discovery of neutrons in the cosmic rays by Locher
(1933), have been fully described by Schopper et al. (1967) and
Lockwood (1973). Theoretical studies of their energy spectrum
and spatial distribution were first made by Bethe et al. (1940)
and were subsequently extended by Hess et al. (1961), Lingenfelter (1963a and b) and Newkirk (1963), using neutron diffusion
and transport programs developed for nuclear reactor studies.
These calculations were restricted to neutron energies of less
than about 10 MeV, where neutron scattering was essentially
isotropic - a necessary condition for the applicability of the
diffusion approximation to the transport equation. Constructing
an energy, altitude, latitude and solar cycle dependent neutron
source based on measurements of the variation of cosmic ray
interaction rates in the atmosphere, Lingenfelter (1963b) calculated the neutron leakage current from the atmosphere, shown

in Figure 1, using a multienergy group diffusion program. Above
10 MeV the neutron energy spectrum was simply extrapolated,
assuming the same shape as that measured by Hess et al. (1959)
in the atmosphere - the only measurement then available.

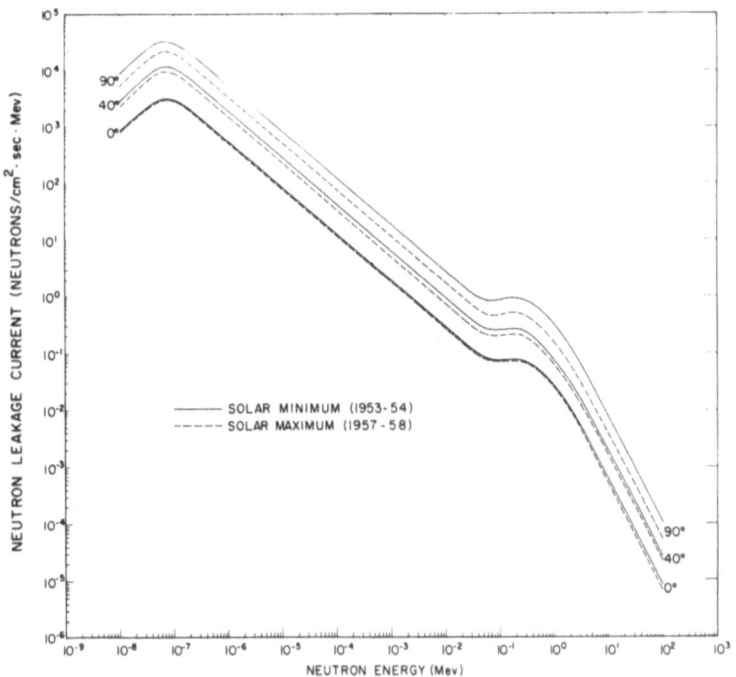

Fig. 1. The cosmic ray neutron leakage current at 0°, 40°
and 90° geomagnetic latitude for solar minimum and solar maximum calculated by Lingenfelter (1963b). The spectral shape
above 10 MeV is extrapolated from the measurements of Hess et
al. (1959).

        Radiation belt models calculated by Dragt et al. (1966)
and Hess and Killeen (1966), using this calculated current as
a source for decay protons and electrons, showed that it was at
least a couple orders of magnitude too small to account for the
observed radiation belt particle intensities. These studies
settled the question for neutrons of energy less than 10 MeV,
since subsequent satellite measurements of the neutron leakage
current in that energy range showed good agreement with the calculated current, as can be seen from the comparison with the
OGO-6 data (Lockwood, 1973) in Figure 2.

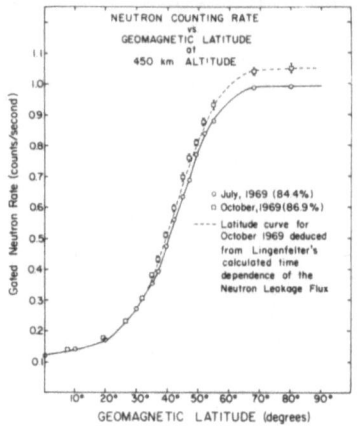

Fig. 2.   The calculated neutron leakage current (<10 MeV) as a
function of latitude compared with that measured on the OGO-6
satellite in July and October 1969 (from Lockwood, 1973).

But at energies greater than 10 MeV there were still no
measurements nor calculations of the neutron leakage current.
Extrapolation from below 10 MeV, however, suggested that this
leakage was also at least an order of magnitude too small to
account for the high energy (>10 MeV) protons in the radiation
belt.

Thus the situation stood until 1972 when Prezler et al.
(1972) made the first detailed measurements of the leakage current with a new instrument sensitive to neutrons in the energy
range from 10 to 100 MeV. They found an intensity more than an
order of magnitude higher than that expected from the simple
extrapolation of the calculated current at lower energies. This
measurement immediately stimulated Monte Carlo calculations of
high energy neutron production and leakage by Merker (1972),
Armstrong et al. (1973) and Light et al. (1973), using the intranuclear cascade model developed by Bertini (1969) and Bertini
and Guthrie (1971). These calculations gave excellent agreement
with the measured leakage current, as can be seen in Figure 3
from White (1973).

Using the measured neutron current Claflin and White (1973)
showed that decay of the 30 to 100 MeV leakage neutrons was indeed sufficient to account for the proton intensity in that same
energy range observed in the radiation belt. In order to make
more detailed calculations of this high energy proton component

of the radiation belt, present experimental and theoretical efforts are directed at a determination of latitude and angular dependence of the high energy neutron leakage current.

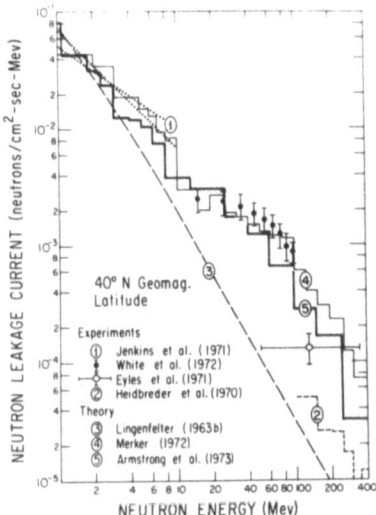

Fig. 3. The measured neutron leakage current at energies >1 MeV, adjusted to 40°N geomagnetic latitude, compared with the current extrapolated from diffusion calculations (Lingenfelter, 1963b) and calculated directly by Monte Carlo programs (Merker, 1972 and Armstrong et al., 1973).

Although most of the radiation belt particles are not produced by decay of cosmic ray neutrons leaking out of the atmosphere as was first proposed, it is now clear that one important component, the high energy radiation belt protons, are produced by this process. Further study of this component can greatly aid in the understanding of such general radiation belt processes as particle interactions with the atmosphere and radial diffusion within the magnetosphere, which are common to all radiation belt particles.

We turn now to the problem of C^{14} production by cosmic ray produced neutrons in the atmosphere. Early estimates of the global averaged C^{14} production rate suggested that it was equal to the estimated global averaged C^{14} decay rate (Libby, 1955). This implied that C^{14} was in static equilibrium between production and decay and that the production rate was essentially constant on time scales of the C^{14} decay mean life of $\sim 8 \times 10^3$ years. Subsequent calculations (see e.g. Lingenfelter, 1963a), however, suggested that the present mean production rate of C^{14} was in fact slightly higher than the present mean decay rate and indicated that only a dynamic equilibrium existed.

At the same time measurements (e.g. Suess, 1971) of C^{14}
activity in samples of known age - at first scattered Egyptian
artifacts and later continuously dated tree rings (Ferguson,
1971) - showed significant variations of as much as $\sim 10\%$ in C^{14}
activity compared to that expected for constant C^{14} production.
Elsasser et al. (1956) had previously suggested that long term
variations in the geomagnetic field, which modulates the global
averaged cosmic ray intensity, could cause variations in the C^{14}
production rate.

This effect has been studied by several authors; the most
detailed study was that performed by Lingenfelter and Ramaty
(1971). They calculated the global averaged C^{14} production as a
function of geomagnetic field strength and solar modulation
(Figure 4). Then using measurements of the variation of the
goemagnetic field over the last ten thousand years, reviewed by
Cox (1969), and assuming average solar cycle modulation (η=-0.7
Gv), they calculated the time dependent C^{14} production rate

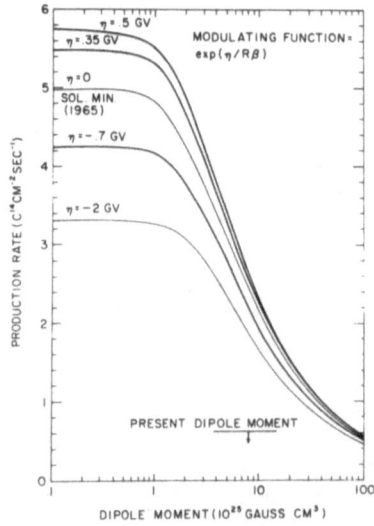

Fig. 4. Global-averaged radiocarbon production as a function of
the geomagnetic dipole moment for various values of the solar
modulation parameter η (from Lingenfelter and Ramaty, 1971).
The solar cycle mean value of η is about -0.7 Gv.

from which they determined the expected variations (ΔC^{14}) in the
biospheric C^{14} activity, using a two reservoir model. These
calculations showed that the bulk of the measured variations in
C^{14} activity could be accounted for by geomagnetic field variations. Subsequent calculations by Dergachev and Kocharov (1972)
using this model and more recent geomagnetic field and C^{14} data
confirm this result. As can be seen in Figure 5, the measured

C^{14} variations can result from a sinusoidal variation in the geomagnetic field strength with a period of $\sim$10,000 years and a ratio (R_O) of the present C^{14} production rate to decay rate of between 1.00 and 1.05.

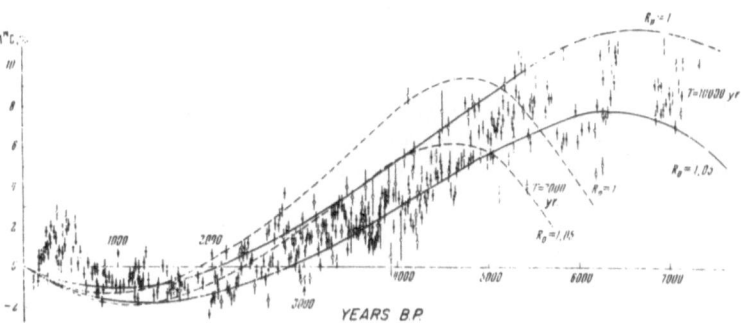

Fig. 5. Comparison of the measured C^{14} variation in tree rings with that expected for a sinusoidal variation of the dipole moment of the geomagnetic field having period τ = 7,000 and 10,000 years and assuming ratios (R_O) between the present C^{14} production and decay rates of 1.0 and 1.05 (from Dergachev and Kocharov, 1972).

If the tree ring chronology can be extended back a few thousand years beyond 7000 B.P. then measurements of the variation of C^{14} activity at these times can greatly improve our knowledge of geomagnetic field variations and also allow us to determine the importance of other possible sources of variation such as long term solar activity varitions.

Shorter term, eleven year solar cycle variations in C^{14} production and activity, are presently being studied. These variations in C^{14} production can result from two competing effects. These are a decrease in C^{14} production resulting from the reduction of the cosmic ray intensity at the Earth caused by stronger, more turbulent interplanetary magnetic fields during periods of maximum solar activity and an increase in C^{14} production resulting from neutrons produced in the atmosphere by solar flare accelerated particles during the same periods of maximum solar activity. Calculations (Lingenfelter and Ramaty, 1971) of the expected variation in the global averaged annual C^{14} production rate due to variations in solar activity over the solar cycle from 1954 to 1965 can be seen in Figure 6. Most of the variation comes from a single large flare on February 23, 1956. Assuming a two reservoir model of the C^{14} equilibrium, this flare produced C^{14} should have caused $\sim$1% increase in the

biospheric C^{14} activity. This increased activity should subsequently have decreased exponentially with a time constant
corresponding to the atmospheric residence time of between 6
and 30 years. Unfortunately comparison of the predicted response
with measurements of C^{14} activity in the atmosphere during that
period is not possible since much larger increases in atmospheric C^{14} were produced by nuclear testing. Studies of possible solar cycle variations of C^{14} activity in tree rings from
earlier periods have recently been reported by Lerman (1971),
Baxter and Walton (1971), Farmer and Baxter (1972) and Damon
et al. (1973a and b). They suggest solar cycle related C^{14}
variations with amplitudes of perhaps ∿0.3%, but larger experimental errors prevent a clear demonstration of such an effect.

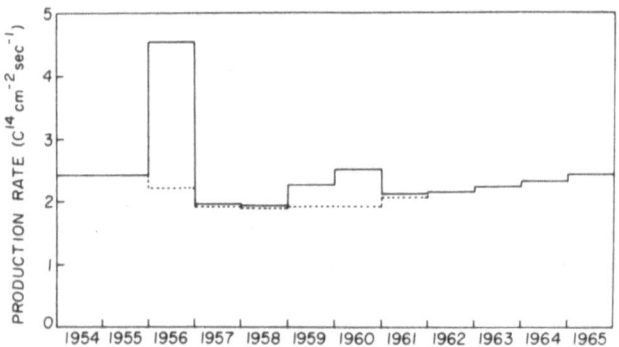

Fig. 6. The variation of global annual average C^{14} production
over solar cycle 19 (from Lingenfelter and Ramaty, 1971). The
decreased radiocarbon production around the maximum of solar
activity (dashed line) resulting from cosmic ray modulation by
the interplanetary magnetic field is obscured by enhanced production due to solar flare accelerated particles.

     Another possible source of short term increases in atmos-
pheric C^{14} is gamma ray bursts from supernovae. Konstantinov
and Kocharov (1965 and 1967) suggested that if supernovae produced as much as ∿10^{50} ergs in high energy (>25 MeV) gamma rays,
then historical supernovae such as Kepler's in 1604 AD could
give an increase in biospheric C^{14} activity of as much as ∿5%,
if the supernova is only 1 kpc away. Such an increase, like
that expected from a solar flare, should then decrease exponentially in time. Subsequent measurements by Sementsov et al.
(1971) of ΔC^{14} activity in tree rings for the years 1593 to
1615, however, showed no increase associated with the supernova
to within the ±0.4% uncertainty of the measurements. Based on
this measurement Dergachev et al. (1971) placed an upper limit

of $\leq 5 \times 10^{48}$ ergs for the total energy in 25 MeV gamma rays produced by this supernova. Kocharov et al. (1974) have also looked for possible C^{14} activity increases in tree rings for the years 1564 to 1583 and 1688 to 1712, which might have been associated with Tycho's supernova of 1572 AD or the Cassiopeia supernova of about 1700 AD, but they found no increase in C^{14} activity $>0.3\%$. Despite these negative results a search should still be made for a possible increase associated with the apparently more energetic Crab Nebula supernova of 1054 AD.

The ~ 7000 year time span of C^{14} activity measurements is too short to search for possible increases in the local cosmic ray intensity, which might be produced by cosmic rays made in a nearby supernova. But the longer-lived isotopies Be^{10} and Al^{26} with decay mean lives of 2.3×10^6 years and 1.1×10^6 years could serve as effective monitors of longer term cosmic ray intensity, since these radionuclides are also produced by cosmic ray spallation of atmospheric nitrogen and argon and are deposited in the deep sea sediments much as C^{14} is in tree rings. Peters (1955 and 1957) first suggested the study of Be^{10} activity in deep sea sediments as a cosmic ray monitor. The idea was advocated here at the University of Pennsylvania by Korff and Mendell in 1967 and I'd like to do the same here again. I believe that such a study is in fact the most promising direction for future research in cosmic ray produced radionuclides. I would like to briefly discuss what we might expect to observe, what has already been done and what can be done.

Based on present estimates of the frequency of supernovae in the galaxy, about 1 every 30 years, of the total energy in cosmic rays which might be produced in a supernova, $\sim 10^{50}$ ergs, and of the rate of propagation of cosmic rays in the interstellar medium, we could expect to see an increase of more than 100% in the local cosmic ray intensity every $\sim 10^6$ years resulting from nearby (<100 pc) supernovae, if they are in fact cosmic ray sources. Such increases might persist for times of the order of $\sim 10^5$ years and should be observed as comparable increases in the Be^{10} and Al^{26} activity in deep sea sediments.

In addition, shorter duration increases in the global averaged cosmic ray intensity might also be expected if the dipole component of the geomagnetic field goes essentially through zero during reversals of polarity. At least two dozen reversals of the geomagnetic field are known over the last 4.5 million years from paleomagnetic studies of continental lavas and oceanic basalts (Cox, 1969). Global averaged cosmic ray increases associated with such reversals could be as large as $\sim 100\%$ and last on the order of $\sim 10^3$ years.

Thus measurements of Be^{10} and Al^{26} activity in deep sea

sediments could contribute significantly to our knowledge of
both astrophysical and geophysical processes.

At the present, however, only one set of measurements
(Amin et al., 1966) has been made of Be^{10} and Al^{26} activities
in two sediment cores from the Pacific Ocean. These measurements were directed at trying to determine the sedimentation
rate from the depth dependence of Be^{10} and Al^{26} activity in the
cores, assuming constant production by the cosmic rays in time.
The Be^{10} and Al^{26} activities, however, were not found to decrease exponentially with depth in the sediment as was expected
and the study was not pursued.

Since then sedimentation rates have been determined by a
number of independent techniques; the most useful of which, I
believe, is by determining the sediment depth of known geomagnetic reversals from measurements of the remnant magnetism in sediment cores. This also permits the determination of possible
variations in the sedimentation rates which is crucial in interpreting the Be^{10} and Al^{26} activities since they are measured with
respect to the amount of sediment.

Measurements of remnant magnetism have not been made on the
two cores studied for Be^{10} and Al^{26} activity by Amin et al.
(1966) but sedimentation rates for other cores in the same regions have been found by this and other techniques to be between
about 3×10^{-5} and 6×10^{-5} cm per year. Assuming such rates for
the two cores, Higdon and Lingenfelter (1973) calculated the
depth and hence time dependence of Be^{10} activity corrected for
decay by the factor $\exp(d/s\tau)$ where d is the depth in the sediment; s is the sedimentation rate and τ is the Be^{10} decay mean
life of 2.3×10^6 years. If the sedimentation rate is constant,
this function is proportional to the global-averaged cosmic ray
intensity at the earth and hence would be a constant with depth
if the cosmic ray intensity were constant in time. The results
of the calculation are shown in Figure 7 for the assumed sedimentation rates of 3×10^{-5} and 6×10^{-5} cm per year.

The anomalous increase at ~ 1.5 and ~ 3 million years ago
also shows up in a similar treatment of the Al^{26} activity. Its
magnitude and duration seem to rule out the possibility that it
might be caused by a reversal of the geomagnetic field. Both,
however, would be consistent with increases which might be
caused by nearby supernovae. The time of the apparent cosmic
ray increase corresponds to the estimated time of 1.6 to 3.3
million years of the supernova explosion which produced the
closest (~ 60 pc) pulsar PSR 1929. The curves in Figure 7 represent the local variation in the cosmic ray intensity that might
have resulted from that supernova if it produced $\sim 10^{50}$ ergs of
cosmic rays, assuming diffusive cosmic ray propagation with a

mean free path in the interstellar medium consistent with measured upper limits on the cosmic ray anisotropy.

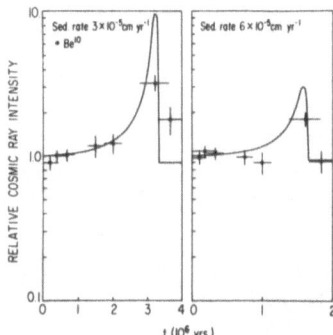

Fig. 7.   The apparent variation of the global averaged cosmic ray intensity as a function of time, deduced from Be^{10} activity measured (Amin et al., 1966) in deep sea sediment cores, assuming sedimentation rates of 3×10^{-5} and 6×10^{-5} cm per year.   The curves represent the intensity increase which might be expected from a nearby supernova explosion (from Higdon and Lingenfelter, 1973).

Although the apparent cosmic ray increase is quite consistent with such an event, this is clearly a purely speculative identification.   Obviously further measurements of Be^{10} and Al^{26} activity in other sediment cores are required to show whether the increased activity is in fact even reproducible and these should be undertaken on cores for which remnant magnetism has also been measured in order to determine whether such an increase could be caused by variations in the sedimentation rate.

Nonetheless, I believe these first measurements by Amin et al. (1966) have clearly shown the great potential which such measurements hold for determining the time history of the cosmic ray intensity at the earth with all its potential biological implications, for furthering our understanding of cosmic ray origin and propagation, and for shedding new light on geomagnetic field variations and possible long term variations in solar activity.

ACKNOWLEDGMENTS

This work was supported in part by the National Science

Foundation under Grant NSF GP 31620.

REFERENCES

Amin, B.S., D.P. Kharkar and D. Lal, 1966, Deep Sea Res., 13, 805.

Armstrong, T.W., K.C. Chandler and J. Barish, 1973, J. Geophys. Res., 78, 2715.

Baxter, M.S. and A. Walton, 1971, Proc. Roy. Soc. London, Ser. A, 321, 105.

Bertini, H.W., 1969, Phys. Rev., 188, 1711.

Bertini, H. W. and M.P. Guthrie, 1971, Nucl. Phys., A169, 670.

Bethe, H.A., S.A. Korff and G. Placzek, 1940, Phys. Rev., 57, 573.

Claflin, E. and R.S. White, 1973, J. Geophys. Res., 78, 4675.

Cox, A., 1969, Science, 163, 237.

Damon, P.E., A. Long and E.I. Wallick, 1973a, Earth and Plan. Sci. Lett., 20, 300.

Damon, P.E., A. Long and E.I. Wallick, 1973b, Earth and Plan. Sci. Lett., 20, 311.

Daniel, R. R. and S.A. Stephens, 1974, Rev. Geophys. Space Phys., 12, 223.

Dergachev, V.A., G.E. Kocharov and Kh.P. Vankov, 1971, Radio-uglerod (Vil´nyus) 49.

Dergachev, V.A. and G.E. Kocharov, 1972, Izv. Akad. Nauk SSSR, ser. fizo, 36, 2312.

Dragt, A.J., M.M. Austin and R.S. White, 1966, J. Geophys. Res., 71, 1293.

Elsasser, W., E.P. Neg and J. R. Winckler, 1956, Nature, 178, 1226.

Eyles, C.J., A.D. Linney and G.K. Rochester, 1971, Proc. Twelfth Internat. Conf. on Cosmic Rays, 2, 462.

Farmer, J.G. and M.S. Baxter, 1972, Proc. Int. Conf. Radiocarbon Dating, 1, 75.

Ferguson, C.W., 1971 in Nobel Symposium 12: Radiocarbon Variations and Absolute Chronology, ed. I.U. Olsson, (Stockholm: Almqvist and Wiksell), 237.

Haymes, R.C., 1965, Rev. Geophys., 3, 345.

Heidbreder, E., K. Pinkau, C. Reppin and V. Schonfelder, 1970, J. Geophys. Res., 75, 6347.

Hess, W.N., 1959, Phys. Rev. Lett., 3, 11.

Hess, W.N., E.H. Canfield and R.E. Lingenfelter, 1961, J. Geophys. Res., 66, 665.

Hess, W.N. and J. Killeen, 1966, J. Geophys. Res., 71, 2799.

Hess, W.N., H.W. Patterson, R. Wallace and E.L. Chapp, 1959, Phys. Rev., 116, 445.

Higdon, J.C. and R.E. Lingenfelter, 1973, Nature, 246, 403.

Jenkins, R.W., S.O. Ifedili, J.A. Lockwood and A. Razdan, 1971, J. Geophys. Res., 76, 7470.

Kellogg, P.J., 1959, Nature, 183, 1295.

Kocharov, G.E., V.A. Dergachev, A.A. Sementsov, E.N. Romanova,
    S.A. Rumyantsev and N.S. Malanova, 1974, Astrofiz.
    Yauleniya; Radiouglerod (Tbilisi) 47.
Konstantinov, B.P. and G.E. Kocharov, 1965, Doklady Akad. Nauk,
    165, 63.
Konstantinov, B.P. and G.E. Kocharov, 1967, A.F. Ioffe Physico-
    Technical Inst., Preprint 064, 43pp.
Korff, S.A. and R.B. Mendell, 1967, in High-Energy Nuclear
    Reactions in Astrophysics, ed. B.S.P. Shen (New York: W.A.
    Benjamin) 159.
Lal, D., and B. Peters, 1967, in Handbuch der Physik, ed. S.
    Flügge (Berlin: Springer-Verlag) 4612, 551.
Lerman, J.C., 1971, in Nobel Symposium 12: Radiocarbon Vari-
    ations and Absolute Chronology, ed. I.U. Olsson (Stockholm:
    Almqvist and Wiksell) 609.
Libby, W.F., 1946, Phys. Rev., 69, 671.
Libby, W.F., 1955, Radiocarbon Dating (Chicago: Univ. of Chicago
    Press).
Light, E.S., M. Merker, H.J. Verschell, R.B. Mendell and S.A.
    Korff, 1973, J. Geophys. Res., 78, 2741.
Lingenfelter, R.E., 1963a, Rev. Geophys., 1, 35.
Lingenfelter, R.E., 1963b, J. Geophys. Res., 68, 5633.
Lingenfelter, R.E. and R. Ramaty, 1971, in Nobel Symposium 12:
    Radiocarbon Variations and Absolute Chronology, ed. I.U.
    Olsson (Stockholm: Almqvist and Wiksell), 513.
Locher, G.L., 1933, Phys. Rev., 44, 779.
Lockwood, J.A., 1973, Space Sci. Rev., 14, 663.
Merker, M., 1972, Phys. Rev. Lett., 29, 1531.
Newkirk, L.L., 1963, J. Geophys. Res., 68, 1825.
Oeschger, H., J. Houtermans, H. Loosli and M. Wahlen, 1971, in
    Nobel Symposium 12: Radiocarbon Variations and Absolute
    Chronology, ed. I.U. Olsson (Stockholm: Almqvist and
    Wiksell), 471.
Peters, B., 1955, Proc. Ind. Acad. Sci., 41, 67.
Peters, B., 1957, Z. Phys., 148, 93.
Prezler, A.M., G.M. Simnett and R. S. White, 1972, Phys. Rev.
    Lett., 28, 982.
Schopper, E., E. Lohrmann and G. Mauck, 1967, in Handbuch der
    Physik, ed. S. Flügge (Berlin: Springer-Verlag) 4612, 372.
Sementsov, A.A., E.N. Romanova, N.S. Malanova, Yu.S. Svezhentsov,
    1971, Radiouglerod (Vil´nyus) 49.
Singer, S.F., 1958, Phys. Rev. Lett., 1, 181.
White, R.S., S. Moon, A.M. Prezler and G.M. Simnett, 1972,
    Space Research, 13, 683.
White, R.S., 1973, Rev. Geophys. Space Phys., 11, 595.

RADIOACTIVE ISOTOPES ON THE MOON[*]

Raymond Davis Jr.

Department of Chemistry, Brookhaven National
Laboratory, Upton, New York

1. EXPECTATIONS

There has been an active interest in observing and accounting
for radioactive products produced by cosmic rays in meteorites
since the early suggestion by Bauer (1947) that the helium
observed in meteorites was mostly produced by cosmic rays. The
first searches for tritium in iron meteorites by Fireman (1958)
led to the possibility of measuring the cosmic ray exposure ages
of meteorites. By the time of the first lunar landing there were
a number of laboratories engaged in measuring radioactive and
stable products produced by cosmic rays in meteorites and earth
bound satellites. These studies led to new knowledge of the
period of time a meteorite fragment was exposed to cosmic rays.
It was found that stone meteorites were exposed for periods of 2
to 30 million years, whereas some iron meteorites were exposed
for many hundreds of millions of years. By studying selected
radioisotope pairs it was found that the galactic cosmic ray
intensity has been essentially constant for the last 500,000
years (^{39}Ar, ^{36}Cl), and perhaps there has been a lower cosmic ray
intensity a few billion years ago (^{36}Cl-^{40}K). Radioactive products were also applied to studying the gradient in the galactic
cosmic ray intensity within the solar system as a result of
modulation by the magnetic fields associated with the solar wind.
Many studies were made of the distribution of cosmic ray spallation products in meteorites and these data were interpreted in
terms of nuclear cascade processes. Information on these various

---

[*] This work supported by the U.S. Energy Research and Development
Administration.

Shen/Merker (eds.), Spallation Nuclear Reactions and Their Applications, 207–223. All Rights Reserved.
Copyright © 1976 by D. Reidel Publishing Company, Dordrecht-Holland.

topics can be obtained from reviews and numerous original papers [1, 2].

We may ask what new information could be obtained by studying radioactive products in lunar samples? Having a man on the moon made it possible to collect particular samples chosen for the problem of interest. In the case of meteorites the original surface is lost and the location of the sample within the pre-atmospheric-impact body is unknown. On the moon's surface, samples can be collected for studying radioactivity distribution in the first few millimeters of the surface. From these surface samples one could learn about solar flare particle fluxes and derive information about the variations of these particle fluxes in the past. Having the surface material one could search for radioactive products collected from space, for example, supernova debris. By digging or boring holes in the lunar soil, samples could be obtained at depth, so that we could derive information about the nuclear cascade processes produced by cosmic rays in a dense medium. This information could be applied to an under-standing of mixing processes in the lunar soil. Exposure age dating of rocks could be useful in determining the history of cratering processes on the moon. It was generally anticipated that a wide range of exposure ages would be obtained. There was hope that from a study of certain radioactive products that might be present in the lunar atmosphere, e.g., tritium and radon, one could learn whether these isotopes were desorbed or vented from the lunar soil or perhaps were injected into the lunar atmosphere by solar wind or solar flare particles. Direct counting of lunar rocks could also measure the primordial radioisotopes ^{40}K, ^{238}U, ^{230}U, and ^{232}Th. Using this technique a survey of these primor-dial radioactive isotopes can be made without sacrificing material. These are some of the problems that could be studied. There was the further vague hope that by entering this new area of research something new would turn up.

It is interesting to recount how the lunar program was operated, what sort of samples were obtained, and what new infor-mation resulted from studying radioactivities in lunar rocks and soil. This report will not be a proper review, but will give only the results of a few studies of special interest from which we can see the broad scope of the program.

2. LUNAR SAMPLES, DISTRIBUTION AND TECHNIQUES

Prior to the first lunar landing the National Aeronautics and Space Administration offered the world scientific community access to lunar samples and funds for studies within the United States. There was a broad response to the offer and appropriate

committees were formed to review the proposals and distribute
lunar material. As a result of six missions to the moon a total
of 381 kilograms of material was recovered. In the early missions
detailed information on sample orientation was not generally
available. As the program developed there was direct communication between the astronauts present on the moon, and a panel of
scientists directing their choice of samples. A broad range of
samples was returned consisting of individual rocks, rock chips,
small rocks raked from the surface, surface soil, soil from
trenches, and core and drill tubes driven into the lunar surface.
The material was transported in sealed rock boxes and the larger
rocks were carried in pouches in the spacecraft. There was a
preliminary examination of selected material that gave useful
data on chemical composition, mineralogical content, and rare gas
content. The results of the preliminary examination for each
mission were published in Science. A description of each rock
and soil sample was given in a rock catalog for each mission.
This basic information was available to the scientist, but in
general was insufficient to select the best sample for a particular purpose. This was especially true for those studying radioactivities where rock orientation and clean surface samples are
needed. Obtaining the most significant samples required a good
working arrangement between the individual investigator and a
member of the Lunar Sample Planning Team. This arrangement was
workable but it was difficult for an investigator to discover
what samples were available and obtain them in a short period of
time. In addition to the administrative and documentary difficulties there was the overriding requirement of a biological
quarantine placed on lunar materials. The biological tests
required a period of six weeks before samples could be distributed. The biological quarantine was imposed during the first
three lunar missions, Apollo 11, 12, and 14, and after that the
restriction was lifted.

There were some observations that were time dependent and
required making measurements during the quarantine period. A low-
level counting laboratory was built 50 feet underground in the
Lunar Receiving Laboratory to make radioactivity measurements.
The samples were placed in stainless steel cans, sterilized and
passed to the counting laboratory for direct counting of the
gamma radiation. The ^{37}Ar (35 day half-life) activity measurements were also performed in the Lunar Receiving Laboratory
during the quarantine period. Measurements of the ^{222}Rn (4 day)
present in the gas in the rock boxes were made by extracting the
gas prior to opening the rock boxes. The searches for magnetic
monopoles, and remnant magnetic properties of rocks were also
made during the quarantine period.

Radioactivities were measured by a variety of techniques.

First there was the non-destructive technique of observing the
gamma radiation directly using a pair of large sodium iodide
crystals. Three laboratories were engaged in these measurements,
the Lunar Receiving Laboratory already mentioned, a similar
prototype system at Oak Ridge National Laboratory, and a laboratory
at Battelle-Northwest, Richland, Washington. Although there were
other laboratories, these three made the majority of the direct
counting measurements. Another approach was to dissolve the
sample, make chemical separations of particular elements, and
measure the beta and gamma radiation using a counting technique
designed to have a high selectivity for a particular radioactive
isotope. Another approach was to vacuum melt the sample to
extract rare gases, hydrogen and carbon dioxide. These gases
were separated using gas separation methods and then placed in
proportional counters for measurement.

I would like to give the results of a few selected experiments. These will include some observations made on the first
few millimeters of the surface where the effects of solar flare
particles are important, some measurements on individual rocks,
and some studies of radioactivities produced deep in the lunar
soil by galactic cosmic rays.

3. SOME RESULTS

3.1 At the surface

A group of chemists [3] at the University of California, San
Diego, under the leadership of J. R. Arnold measured the depth
variation of radioactivities in the outer centimeter thick layer
of rock 12002. A vertical slice and a top slab of the rock was
dissected by carefully grinding off layers, see Fig. 1. Analyses
were made by wet chemical methods to study a variety of isotopes,
namely ^{10}Be, ^{22}Na, ^{26}Al, ^{36}Cl, ^{53}Mn, ^{54}Mn, ^{55}Fe, ^{56}Co, and ^{57}Co.
Their results are given in Table 1. The isotope ^{56}Co (77 day
half-life), was produced by (p,n) reaction (threshold 5.4 MeV)
on the abundant isotope, ^{56}Fe (91.7%). Cobalt-56 showed the
steepest profile, and is the best indicator of solar flare
protons. The energy spectrum of solar flare protons is known
reasonably well from satellite data, and the (p,n) cross sections
on ^{56}Fe are well known. Furthermore, the flares of April 12 and
November 3, 1969 contributed 94 percent of the ^{56}Co activity.
The contribution from galactic cosmic rays was essentially
negligible. These investigators compared their ^{56}Co depth profile
from 0.1 to 1.9 g/cm^2 with the depth profile calculated from solar
flare proton intensities and cross-section data. The comparison
for this isotope was excellent. Similar calculations for the

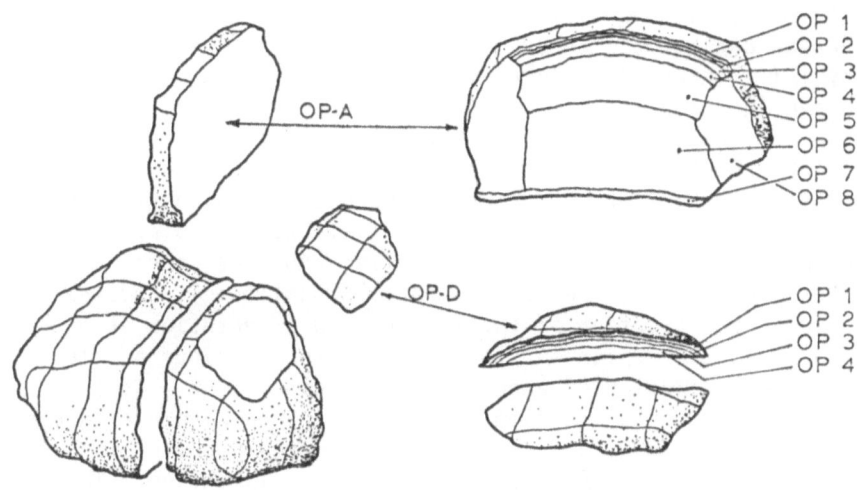

Fig. 1. Schematic of rock 12002, showing location of samples (on left) and subdivision of samples (on right).

Table 1. Activities in rock 12002 (all activities in dpm/kg)

Isotope	$T_{1/2}$	OP-1 9.9 g (0–1 mm)	OP-2 9.5 g (1–2 mm)	OP-3 17.7 g (2–4 mm)	OP-4 25.3 g (4–9 mm)	OP-5 18.2 g (9–20 mm)	OP-6 50.4 g (20–60 mm)	OP-7 4.4 g (60 mm)
Be^{10}	2.5×10^6 y					15.4 ± 3.0	11.5 ± 1.4	
Na^{22}	2.6 y	166 ± 18	122 ± 14	91 ± 11	71 ± 8	54 ± 7	39 ± 6	
Al^{26}	7.4×10^5 y	209 ± 26	154 ± 21	111 ± 16	91 ± 12	72 ± 11	64 ± 6	66 ± 11
Cl^{36}	3.1×10^5 y	13.4 ± 3.8		9.7 ± 2.5	14.7 ± 2.4	8.4 ± 2.1	9.4 ± 1.4	
Mn^{53}	3.7×10^6 y	98 ± 6	86 ± 5	73 ± 5	67 ± 4	52 ± 3	47 ± 3	51 ± 3
Mn^{54}	303 d	98 ± 17	77 ± 12	53 ± 11	50 ± 9	39 ± 10	31 ± 5	
Fe^{55}	2.6 y	731 ± 68	454 ± 46	285 ± 30	171 ± 23	101 ± 17	56 ± 12	
Co^{56}	77 d	533 ± 70	205 ± 30	80 ± 15	32 ± 9			
Co^{57}	270 d	16 ± 6	<6					

longer lived isotopes, ^{54}Mn (303 day), ^{22}Na (2.6 year), and ^{55}Fe (2.6 year) were not nearly so satisfactory, the measured activities were about a factor of two higher than the activity derived from proton fluxes from satellites and cross-section data. The lack of agreement could be attributed to errors in the solar flare particle fluxes, since a large fraction of the total flux was due to large flares from the previous solar cycle and accurate satellite measurements were not yet available at that time.

Of particular interest are the profiles for the longer lived ^{26}Al (7.4 x 10^5 year) and ^{53}Mn (3.7 x 10^6 year). These isotopes can be used to test whether or not the solar flare intensity has varied over the last million years. The usual stellar structure

arguments favor no change in the solar luminosity over the last
10^8 years, although it is entirely possible that major climatic
changes can be attributed to variations in the solar luminosity [4]. It would be interesting to know whether there was a
change in the solar flare intensity, and if so, the depth variation of ^{26}Al and ^{53}Mn may be the only way of testing for this
variation. Unfortunately, the answer is not clear. The measurements on these isotopes agree reasonably well with flare particle
fluxes that are similar in energy distribution and spectral shape
to those of the last two solar cycles. However, there may be
errors in the cross-sections (mostly unpublished). Additionally,
there is the question of the long period orientation of rock
12002, and the degree of space erosion is a serious problem. I
feel the conclusion of essentially constant solar proton
intensity-energy reached is perhaps only true within a factor of
two or so. This is not of adequate accuracy to determine whether
or not there is a solar derived cause for glaciation. This
important question warrants further study, and it would be advis-
able to use the surface of a very large boulder less likely to
have been disturbed.

Another study that could give information about past solar
intensities is the study of the ^{14}C profile. Begeman et al. [5]
measured ^{14}C at three depths (0-0.5; 0.5-2; 2-6.5 cm) in rock
12053 and found the exterior layer was 2.3 times higher than the
two inner layers. This high surface value cannot be explained by
(p,3p) reaction on oxygen with the same average solar proton
intensity and energy spectrum that fits the ^{22}Na, ^{55}Fe, and ^{54}Mn
data, that is, a 4π integral flux above 10 MeV of 100 protons
cm^{-2} sec^{-1}, and an exponential rigidity of 100 MV. To explain
the high ^{14}C value the authors say a flux 3 to 5 times higher,
averaged over the past 10,000 years, is needed. There is also
the mundane possibility that the cross-sections used are incorrect.
A further possibility is that the ^{14}C was produced directly in
the photosphere of the sun and implanted in the surface. A ^{14}C
depth profile was measured by Boeckl [5] on rock 12002, the same
rock studied by the Arnold group (Fig. 1). His measurements were
made with a finer depth structure (4 points 0-9 mm) than those of
Begeman et al. Boeckl obtained a steep profile that was consistent
with a flare proton rigidity that fitted the ^{22}Na, ^{55}Fe, and ^{54}Mn
data, but required a flux a factor of two higher. It is clear
that further studies of the ^{14}C activity in the surface of lunar
rocks is needed to resolve these important questions.

It is known from the historical record that sunspots were
essentially absent over a 70 year period from about 1645 to 1715.
It would be interesting to observe this depression in the solar
activity with radionuclides in lunar surface material. However,
there are no radionuclides with a half-life of a few hundred

years that are suitable for this purpose. Argon-39 (269 y) has the required half-life but is produced primarily by neutron reactions on potassium and calcium. Silicon-32 ($\sim$280 y) is another isotope with the correct half-life, but it is not produced in good yield by solar protons on lunar materials. So it is unfortunate that this solar quiet period cannot be verified quantitatively by radioisotopes, or conversely, that the radioisotope technique be tested!

The surface of the moon is exposed and collects infalling material for billions of years. The temporal record is disturbed in the lunar soil by mixing processes mainly as a result of meteorite impacts. Mixing by impacts and land slides will be serious in some locations but at other locations the soil is relatively undisturbed for long periods of time. There is quantitative information on these processes based upon measurements of the changes in the gadolinium and samarium isotope abundances induced by neutron capture. Studies on an Apollo 15 core allowed Russ et al. [6] to conclude that the upper 10 cm of soil is turned over every 0.1 b.y. and the turnover time at a few meters depth is more than 0.5 b.y.

It is of great interest to search in this material for debris from supernova explosions and for radioisotopes accumulated from the sun. Fields and his associates [7] at Argonne National Laboratory have made a search for ^{244}Pu ($t_{1/2}$ = 82 m.y.) in 12 lunar samples. The lowest limit set was 10^{-17} g ^{244}Pu/g on a few gram sample of an Apollo 15 soil. It is difficult to estimate the expected ^{244}Pu collection rate from supernovas, or cosmic rays. If we take the present upper limit to the cosmic ray intensity for particles of charge 92-96 of 10^{-11} cm^{-2} sec^{-1}, assume 10 per cent of the incoming ^{244}Pu atoms survive stopping in 1 g/cm^2, and collect for a period of 10^8 years, then there may be 10^{-18} g Pu/g of lunar soil. However, if there were low energy ^{244}Pu atoms arriving as supernova debris a higher concentration could be present in lunar soil. It seems clear that further searches for ^{244}Pu should be made with carefully selected samples. There has been a great reluctance to using lunar samples for exploratory work of this nature in spite of the high interest in this problem.

The group at Argonne [7] also made measurements of the ^{236}U/^{238}U ratio (^{236}U $t_{1/2}$ = 23.4 m.y.) and the ^{237}Np ($t_{1/2}$ = 2.14 m.y.) concentration in a number of lunar soils and rock samples. One expects a ^{236}U/^{238}U ratio $\sim$5 x 10^{-9} from slow neutron capture on ^{235}U (total U 0.1 to 2 ppm), and indeed five samples are in this range. However, one rock (12073) had a ratio of 40-50 x 10^{-9} and two samples of soil (12070) had ratios of 30-50 x 10^{-9} (coarse fraction) and 235 x 10^{-9} (total sample). These very high ratios cannot be explained by nuclear reactions on ^{238}U such as the (n,3n) or (p,t) reactions. They require a higher production rate sometime in the past, or some other source for the $236U$. If it were from an increased solar or galactic cosmic ray intensity over long periods of time, all the exposure ages derived from present cosmic ray fluxes would be shortened considerably. However, if there were short periods of very high particle fluxes exposure ages would not be affected. The fact that $53Mn$ activities are within a factor of two of the concentrations expected from present day solar and galactic cosmic ray fluxes requires that fluxes as high as indicated by $236U$ concentrations did not occur during the last 10^7 years. This interesting result needs a clear explanation and it would be very useful to have additional measurements on both rock and soil samples.

3.2 At depth

During the last three missions to the moon the astronauts recovered long columns of the lunar soil with a special coring drill [8]. The length of the three cores were 2.4, 2.4, and 2.9 meters. Obtaining these long cores was a major achievement in sampling the lunar surface. These and the shorter drive tubes are our only samples that give a view of regolith stratigraphy. Studies on these cores using a variety of techniques, gadolinium and samarium isotope mass spectrometry, rare gas isotope analyses, fossil track distributions, chemical analysis, and minerological studies have lead to new knowledge of the history of the lunar regolith. I would like to confine my discussion to an in situ measurement of the neutron flux distribution and some radio-isotope measurements that were made on these cores.

A basic factor in understanding the gadolinium and samarium isotope distribution changes as a result of neutron capture is to know the present-day neutron fluxes produced in the lunar soil by cosmic ray interactions. The flux distribution was calculated theoretically by Lingenfelter, Canfield, and Hampel and by Armstrong and Alsmiller [9], and it was of great interest to measure the flux experimentally. The measurement was made by a special probe designed by Don Burnett (CIT) that was introduced into the hole in the lunar regolith after the Apollo 17 coring drill was removed. After insertion a mica fission track detector was rotated over a $235U$ foil, and a plastic track detector was rotated over a $10B$ foil to register slow neutron captures. The probe was in place registering the flux for 49 hours. The results [10] for both target elements showed a sharp rise from the surface to a broad peak at a depth of 100-160 g/cm^2 and slow decrease with increasing depth. The shape of the curve fitted the theoretical curve very well, see Fig. 2. The total neutron capture rate in $235U$ was 11 $\pm$ 17 percent lower, and in $10B$ it was 19 $\pm$ 13 lower than the theory of Lingenfelter et al. predicted.

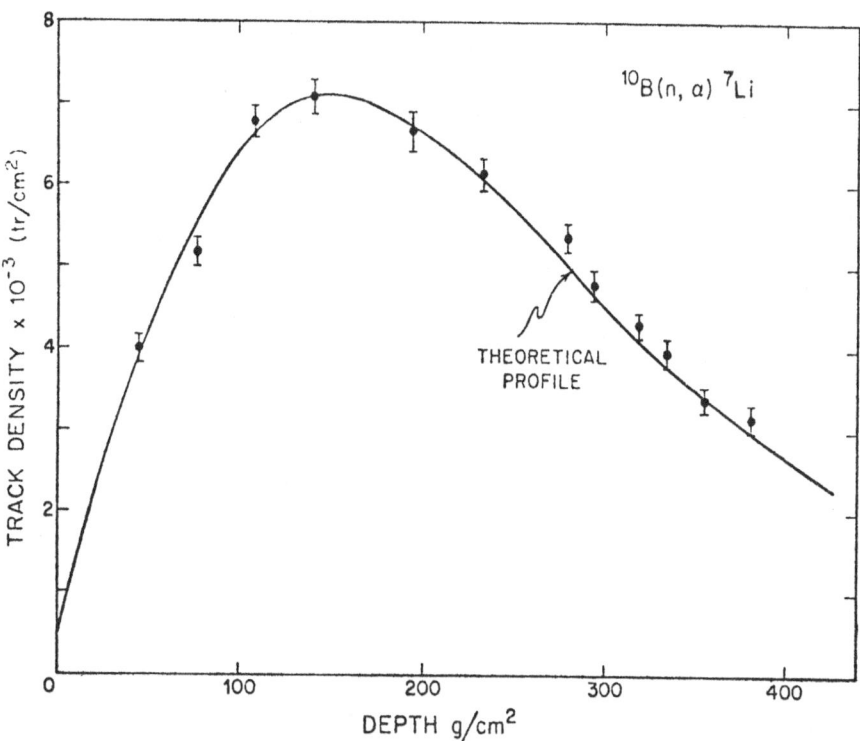

Fig. 2. Neutron depth profile obtained by Woolum and Burnett [10] compared to the theory of Lingenfelter et al. [9].

The results for ^{10}B normalized to the theory are shown in Fig. 2.

By combining the neutron flux distribution in the lunar soil with measurements of the total neutron fluences (flux X time) measured by ^{157}Gd captures and the regolith depth (4-5 m) inferred from observations, it has been concluded that the total fluence is a factor of 2-3 times lower than expected for a uniformly mixed soil exposed for 3.8 billion years. Workers in this field have made a great issue of this problem [10, 11]. The various suggested solutions include (1) the regolith is not well mixed, (2) the regolith is not as deep as is thought, (3) the cosmic ray intensity averaged over the last 3-4 billion years is lower than present intensities, (4) large quantities of lunar soil are ejected by meteroil impacts [12] (40-100 cm/10^9 y), or (5) some combination of these or other discrepancies give an apparent reduction in neutron fluences. These observations give an interesting insight into the development of the lunar regolith, a topic of broad significance in the evolution of the solar system.

It was of interest to have a measure of the fast neutron flux generated in the soil by cosmic rays. The 35-day isotope ^{37}Ar is

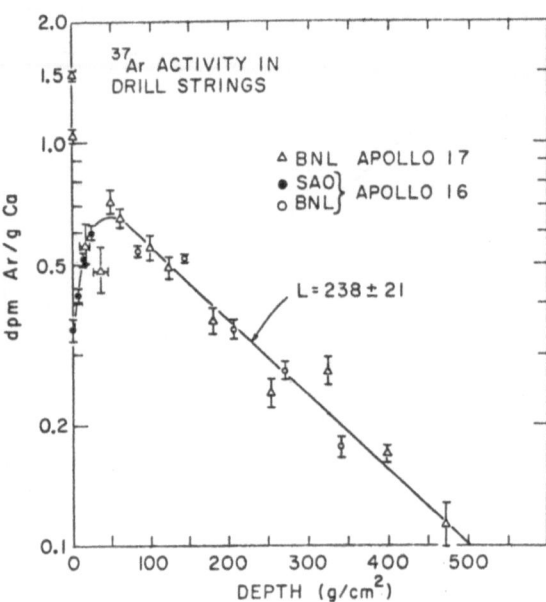

Fig. 3. ^{37}Ar depth profile measured in the Apollo 16 and 17 drill cores [14, 15].

an ideal one for this purpose. It is produced by the reaction ^{40}Ca(n,α)^{37}Ar on the abundant element calcium (7-12 percent). Argon-37 is a radioisotope that is easily extracted and counted making it possible to perform measurements on modest amounts of sample (1-2 g). The production cross section has been measured. The excitation function shows an effective threshold at about 3 MeV, rising to broad maximum at 6 MeV (205 mb) and then it drops slowly with increasing energy (138 mb at 14 MeV) [13]. The ^{37}Ar activity as a function of the depth was measured in the Apollo 16 core as a joint project between Smithsonian Astro-physical Observatory [14] and Brookhaven National Laboratory [15]. Only the deeper portions of the Apollo 17 core were analyzed because the upper portions of this core were not opened in time for the measurements. All samples were analyzed for calcium so that the measurements for the two missions could be compared. A plot of dpm ^{37}Ar/g Ca is given in Fig. 3. The ^{37}Ar production exhibits a steep rise with depth, corresponding to the development of the nuclear cascade, reaching a maximum at a depth of about 50 g/cm^2, and then it dropped exponentially with an attenuation length of 240 $\pm$ 20 g/cm^2. This general behavior is expected from the nuclear cascade process in the lunar soil [16]. There were no solar flares immediately before the Apollo 16 mission, so these measurements give a valid ^{37}Ar production rate from galactic

cosmic rays. Kornblum and his associates [17] have made calculations of the ^{37}Ar depth profile with a defined neutron source
function, the neutron moderating factors derived from the lunar
composition (ANISN neutron transport code), and the measured
^{40}Ca$(n,\alpha)^{37}$Ar excitation function. Their calculated profile fits
the measurements, and they obtain a total neutron production rate
in lunar soil of 26 ± 4 cm^{-2} sec^{-1} by normalizing the calculations
to the experimental measurements. This value is higher than that
of Lingenfelter et al. and Armstrong and Alsmiller, 16 ± 5 and
17.5, respectively.

The Apollo 17 mission followed only 124 days after the
intense solar flare of Aug. 4-9, 1972. This flare was the most
intense flare ever observed, similar in magnitude to the great
flare of 1956. This flare produced approximately 2×10^{10}
protons/cm^2 with energy above 60 MeV. The effects of this flare
were clearly observed in many samples returned from the Apollo 17
mission. The ^{37}Ar activities were found to be very high in the
samples from the surface (depth 0-5 g/cm^2), but samples at greater
depths (>50 g/cm^2) were not affected. The flare protons interact
at the surface to produce ^{37}Ar by the reaction ^{40}Ca$(p,\alpha)^{37}$K$(\beta^+$-
decay)^{37}Ar. During the Aug. 1972 flare the production rate at
the surface was over 400 times the normal cosmic ray production
rate.

The long lived argon isotope, ^{39}Ar $(t_{1/2} = 265$ y$)$, was also
measured on the Apollo 16 and 17 cores [14, 15]. This isotope is
produced by several processes, ^{39}K$(n,p)^{39}$Ar, ^{40}Ca$(n,2p)^{39}$Ar, and
by spallation of iron. The excitation functions for the neutron
reactions have not been measured so the relative cosmic ray
production from these various processes cannot be sorted out.
The depth distribution curve exhibits the characteristic increase
from surface values (factor of 1.6) reaching a maximum at a depth
of 50 g/cm^2 and then dropping exponentially with an attenuation
length of 166 g/cm^2. The general behavior is in agreement with
expectation.

It is of great interest to observe the depth profile of a
long lived radioactive isotope to search for long period variations in solar activity or galactic cosmic ray intensities. As
mentioned earlier ^{53}Mn is an ideal isotope for this purpose. It
has a long half-life, 3.7×10^6 years, and it can be measured by
a neutron activation technique that requires only a 0.2 g sample.
This isotope is made at the surface by low energy protons mainly
by the ^{56}Fe$(p,\alpha)^{53}$Mn and ^{56}Fe$(n,p3n)^{53}$Mn. Imamura et al. [18]
measured the ^{53}Mn depth profiles in the Apollo 15 and 16 cores.
The solar proton produced ^{53}Mn is very high at the surface,
dropping rapidly with depth to a depth of 2 g/cm^2, see Fig. 4.
Below a depth of 10 g/cm^2 it decreases exponentially with an
attentuation length of 220 ± 50 g/cm^2. The results agree well

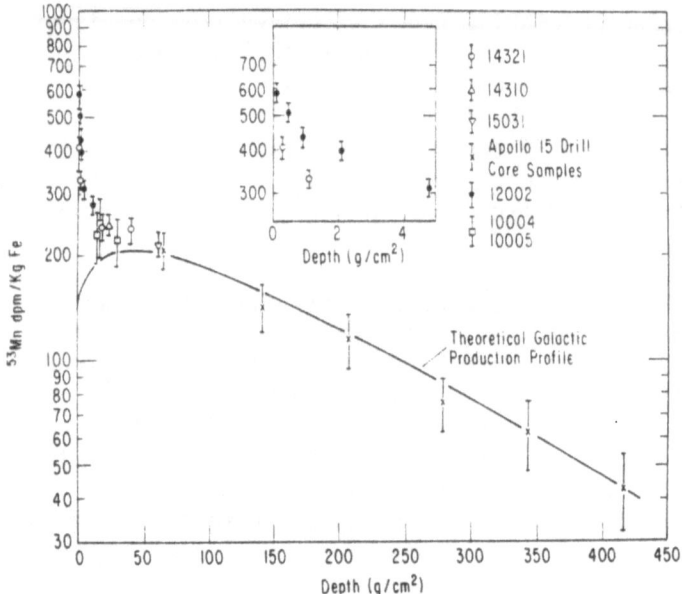

Fig. 4. ^{53}Mn depth profile of imamura et al. obtained on the Apollo 15 drill core, soil and rock samples [3, 18].

with the model calculations of Reedy and Arnold [16]. It was concluded that the soil column has remained undisturbed for the last 5 million years, consistent with the Gd isotope conclusions mentioned earlier that extend over a much longer period of time.

It would be interesting to compare the depth profile of various radioisotopes with actual measurements with accelerated protons and alpha particles. In the past there has been a number of these thick target bombardments, the first being B. S. P. Shen's bombardment of an iron target 30 x 30 x 90 cm with 3 GeV protons. The results of these early bombardments are summarized in a review by Kohman and Bender [2]. Two studies have been made recently with thick targets of simulated lunar soil. Bar and Herr [19] studied thermoluminescence of minerals as a function of the depth in a target 30 x 30 x 140 cm bombarded with 600 MeV protons. Lyman [20] measured ^{37}Ar, ^{39}Ar, ^{42}Ar, and tritium activity distribution in a simulated lunar target 20 x 20 x 43 cm bombarded by 600 MeV protons. Experiments of this type require counting numerous samples and it is further necessary to perform the measurements at a series of energies. In view of the enormous effort involved, most experimenters are discouraged. Also the basic nuclear cascade process is understood, and it does not seem worth the effort at the present time to establish the detailed parameters involved.

3.3 Individual rocks

The largest amount of data on radioactivities in lunar
materials was obtained by a few laboratories that measured gamma
radiation by NaI scintillation counters. Large pure activated
NaI(Tl) crystal scintillation detectors were developed industrially
and a number of laboratories devised sophisticated low-level
counting arrangements for measuring lunar and environmental
samples. This approach has the decided advantage of being able
to measure lunar samples without affecting them in any way. Once
the elaborate counting facility was built a large number of
samples can be surveyed. In fact the accumulation of data was
limited by the ability of the curator's office to process the
sample, especially during the quarantine period. One extremely
useful observation that was made immediately was that lunar rocks
are relatively low in the volatile element potassium. If a plot
of the ratio K/U versus K concentration is made, it can be shown
that igneous earth rocks, lunar rocks, and various meteorite
classes, eucrites, carbonaceous chondrites and ordinary chondrites
fall into distinct groups [21]. This useful plot not only sorts
out solar system materials but also allows a classification of
lunar basalts, soils, highland rocks, and breccias. These
measurements of primordial radioactive elements also demonstrated
that the Th/U ratio for lunar rocks is 3.8, almost identical to
the ratio generally observed for earth rocks [21, 22].

The usual cosmogenic radioactivities measured were: ^{22}Na
(2.6 y), ^{26}Al (3.7 x 10^6 y), ^{46}Sc (84 d), ^{48}V (16 d), ^{51}Cr (28 d),
^{54}Mn (303 d), ^{57}Co (270 d), ^{58}Co (71 d) and ^{60}Co (5.3 y). The
activity levels are dependent upon the chemical composition of
rock, and where relevant cross-sections and chemical analyses
are known, the agreement between measured and calculated activities is satisfactory. There is interest in deriving information
about cosmic ray intensities in the past and about lunar surface
events from the pair of isotopes ^{22}Na and ^{27}Al. These isotopes
are produced by various reactions on the elements Si, Al, Mg, and
Na. Yokayama et al. [23] and Keith and Clark [24] have attempted
to normalize the observed activities by adopting a production
rate for each element. Their techniques are useful for selecting
rocks with exposure ages comparable to the half-life of ^{26}Al, the
so-called undersaturated rocks.

The solar flare of Aug. 4-9, 1972 produced dramatic increases
in activity levels of all measured isotopes with half-lives
shorter than a few years. By careful measurements of various
isotopes with well measured excitation functions it was possible
to determine the average rigidity parameters and fluxes that

characterize the energy spectrum of the solar flare protons from
this event [21, 25]. Also there is evidence from the measurements
on rock surfaces at various angles of inclination that the flare
particles were anisotropic [25].

I might conclude by some remarks about cosmic ray exposure
ages. Since lunar rocks are exposed on the surface, one can apply
mass-spectroscopic measurements of the abundant rare gas spallation products like ^{3}He or ^{38}Ar to derive exposure ages, assuming
a constant production rate for all samples. Helium-3 exposure
ages are questionable because of the loss of both tritium and
helium. A much more reliable method is to determine the temperature release pattern of the argon isotopes from neutron activated
samples following the technique now common in ^{40}Ar-^{39}Ar dating.
In this method the release of spallation ^{38}Ar from calcium can be
related to the ^{37}Ar produced by (n,α) reaction on ^{40}Ca. By
adopting an assumed ^{38}Ar cosmic ray spallation production rate on
calcium (1.4×10^{-8} cm^3 STP/g Ca$\cdot 10^6$ y), the exposure age can be
obtained. The usual assumption of constant spallation production
for all samples is of course not correct, as there is a change in
production rate with depth. Another method of determining exposure ages depends on mass spectroscopic measurements of ^{81}Kr
($t_{1/2} = 2.1 \times 10^5$ y), a spallation product of Sr, Y, and Zr. The
method depends upon the rock being exposed for a period long
compared to the half-life of ^{81}Kr permitting the concentration of
^{81}Kr to be used for the production rate in the sample. The concentration of ^{81}Kr relative to the stable isotope ^{83}Kr is used to
determine the exposure age. It is assumed that the production
cross section of ^{81}Kr is equal to the average of the ^{80}Kr and ^{82}Kr
production cross-sections. One would expect the relative production cross-sections of ^{81}Kr and ^{83}Kr to depend upon target element,
and thereby depend upon the Sr, Y, and Zr composition of the rock.
However, in spite of this obvious difficulty, this method is held
in high esteem.

Exposure ages of lunar rocks vary considerably, as one might
expect from random impacts on the lunar surface so precise
exposure age measurements are not particularly required. There
are numerous exposure age measurements of variable quality
throughout the lunar literature. The range of exposure ages are
from 600 million years to rocks that have not been exposed long
enough for ^{26}Al activity to reach saturation, about 1 million
years.

4. CONCLUSIONS

It is interesting to look back over the accomplishments of
the last five years of studying radioactivities in lunar samples,
and to inquire whether we have learned anything new. There is no

question that an enormous amount of new information has resulted from studying lunar samples and we can address our inquiry at three levels: (1) were the new observations consistent with or do they support earlier views, or (2) were there some surprises that now need confirmation, or (3) did we discover something totally new. I believe the studies of radioactivities in lunar material are essentially characterized as being consistent with earlier knowledge, but there are a few interesting results that need further work.

First, concerning the production of radioactivities by galactic cosmic rays. The measurements in individual lunar rocks and at depth in the lunar soil are consistent with what was already known. The measurements of the depth profile of radioactive products with half-lives in the order of months to millions of years agree reasonably well with theoretical calculations of the nuclear cascade processes and present galactic cosmic ray intensities and energy spectra. The lunar results are consistent with a constant cosmic ray flux. However, the best test of the constancy of the galactic cosmic ray flux are the older measurements of ^{39}Ar, ^{36}Cl, and ^{40}K in iron meteorites. It seems unlikely that further measurements on lunar material will lead to any new knowledge on this question. However, one can see that there will be further interest in applying our knowledge of cosmic ray interactions in lunar material to understanding mixing processes in the lunar soil, and possibly measuring accurate exposure ages of lunar rocks.

Studies of the activities on the surface of lunar rocks and surface soil samples have produced some interesting results. The question of the solar flare intensity in the last 10^6 or 10^8 years is an important question that needs further study. New measurements of the depth profile are needed of ^{22}Na, ^{14}C and ^{53}Mn on the surface of large rocks likely to be fixed in their position for periods longer than 10^7 years. Additional studies of ^{236}U and ^{237}Np are certainly needed. It is very important to understand how some samples can have 10 to 100 times higher concentrations of ^{236}U than others. Perhaps ^{236}U is the only isotope with a half-life long enough to retain the effects of events that occurred 10^8 years ago.

Finally a thought on the use of lunar samples for exploratory work. Experiments like the ones searching for ^{244}Pu, and super heavy elements have not been especially encouraged. It may be a few decades before new samples are obtained from the moon, but when lunar exploration is revived our present closely guarded collection will be only of secondary interest. I feel the present collection should be more available to the scientific community and more emphasis should be placed on significant exploratory studies.

REFERENCES

The abbreviation PLSC-2 stands for Proc. 2d Lunar Sci. Conf.,
Geochim. Cosmochim. Acta, Suppl. 2.

1.   O. A. Schaeffer, Ann. Rev. Phys. Chem. 13, 151 (1962); M.
     Honda and J. R. Arnold, Handbuch der Physik XLV1/2, p. 613
     (1967), Springer-Verlag, Berlin.
2.   T. P. Kohman and M. L. Bender, High-Energy Nuclear Reactions
     in Astrophysics, ed. by B. S. P. Shen, Benjamin Press, 1967.
3.   R. C. Finkel, J. R. Arnold, M. Imamura, R. C. Reedy, J. S.
     Fruchter, H. H. Loosli, J. C. Evans, and A. C. Delany,
     PLSC-2, Vol. 2, p. 1773 (1971).
4.   A. G. W. Cameron, Rev. Geophys. and Space Physics 11, 505
     (1973).
5.   F. Begeman, W. Born, H. Palme, E. Vilcsek, and H. Wanke,
     PLSC-3, Vol. 2, p. 1693 (1972); R. S. Boeckl, Earth and
     Planetary Sci. Letters 16, 269 (1972).
6.   G. P. Russ III, D. S. Burnett, and G. J. Wasserburg, Earth
     and Planetary Sci. Letters 15, 172 (1972).
7.   P. R. Fields, H. Diamond, D. N. Metta, and D. J. Rokop,
     PLSC-4, Vol. 2, p. 2123 (1973), contains earlier references.
8.   D. S. Woolum, D. S. Burnett, and C. A. Bauman, Apollo 17
     Preliminary Science Report NASA SP-330. These reports are a
     very useful source of general information on each mission.
9.   R. E. Lingenfelter, E. H. Canfield, and V. E. Hampel, Earth
     and Planetary Sci. Letters 16, 355 (1972); T. W. Armstrong
     and R. G. Alsmiller, PLSC-2, Vol. 2, p. 1729 (1971).
10. D. S. Woolum and D. S. Burnett, Earth and Planetary Sci.
     Letters 21, 153 (1974); D. S. Burnett and D. S. Woolum,
     PLSC-5, Vol. 2, p. 2061 (1974).
11. O. Eugster, F. Tera, D. S. Burnett, and G. J. Wasserburg,
     Earth and Planetary Sci. Letters 8, 20 (1970); G. Price Russ
     III, ibid, 19, 275 (1973) and references therein.
12. E. L. Fireman, PLSC-5, Vol. 2, p. 2075 (1974).
13. J. W. Barnes, B. P. Bayhurst, B. H. Erkkila, J. S. Gilmore,
     N. Jaimie, and R. J. Prestwood, J. Inorg. Nucl. Chem. 37,
     399 (1975).
14. E. L. Fireman, J. D'Amico, and J. De Felice, PLSC-4, Vol. 2,
     p. 2131 (1973).
15. R. W. Stoenner, R. Davis Jr., E. Norton, and M. Bauer,
     PLSC-5, Vol. 2, p. 2211 (1974).
16. R. C. Reedy and J. R. Arnold, J. Geophys. Res. 77, 537 (1972).
17. J. J. Kornblum, E. L. Fireman, M. Levine, and A. A. Aronson,
     PLSC-4, Vol. 2, p. 2171 (1973); J. J. Kornblum, Thesis,
     State University of New York at Stony Brook, 1975.
18. M. Imamura, R. C. Finkel, and M. Wahlen, Earth and Planetary
     Sci. Letters 20, 107 (1973); M. Imamura, K. Nishiizumi, M.
     Honda, R. C. Finkel, J. R. Arnold, and C. P. Kohl, PLSC-5,
     Vol. 2, p. 2093 (1974).

19.   K. Bar and W. Herr, Earth and Planetary Sci. Letters <u>22</u>, 188
      (1974).
20.   W. J. Lyman, Preliminary Report, Chemistry Department,
      Brookhaven National Laboratory, Upton, N.Y. (1971).
21.   J. S. Eldridge, G. D. O'Kelley, and K. J. Northcutt, PLSC-5,
      Vol. 2, 1025 (1974) and earlier references therein.
22.   J. E. Keith, R. S. Clark, and L. J. Bennett, PLSC-5, Vol. 2,
      2121 (1974) and earlier references therein.
23.   Y. Yokoyama, J. L. Reyss, and F. Guichard, PLSC-5, Vol. 2, p.
      2231 (1974).
24.   J. E. Keith and R. S. Clark, PLSC-5, Vol. 2, p. 2105 (1974).
25.   L. A. Rancitelli, R. W. Perkins, W. D. Felix, and N. A. Wogman,
      PLSC-5, Vol. 2, p. 2185 (1974).

INDEX OF SUBJECTS

INDEX OF NAMES

The page number of the beginning of each chapter is underlined.

ASTROPHYSICS AND SPACE SCIENCE LIBRARY

Edited by

J.E. Blamont, R.L.F. Boyd, L. Goldberg, C. de Jager, Z. Kopal, G.H. Ludwig, R. Lüst,
B.M. McCormac, H.E. Newell, L.I. Sedov, Z. Švestka, and W. de Graaff

24. B.M. McCormac (ed.), *The Radiating Atmosphere. Proceedings of a Symposium Organized by the Summer Advanced Study Institute, held at Queen's University, Kingston, Ontario, August 3–14, 1970.* 1971, XI + 455 pp.

25. G. Fiocco (ed.), *Mesospheric Models and Related Experiments. Proceedings of the 4th ESRIN-ESLAB Symposium, held at Frascati, Italy, July 6–10, 1970.* 1971, VIII + 298 pp.

26. I. Atanasijević, *Selected Exercises in Galactic Astronomy.* 1971, XII + 144 pp.

27. C.J. Macris (ed.), *Physics of the Solar Corona. Proceedings of the NATO Advanced Study Institute on Physics of the Solar Corona, held at Cavouri-Vouliagmeni, Athens, Greece, 6–17 September 1970.* 1971, XII + 345 pp.

28. F. Delobeau, *The Environment of the Earth.* 1971, IX + 113 pp.

29. E.R. Dyer (general ed.), *Solar-Terrestrial Physics/1970. Proceedings of the International Symposium on Solar-Terrestrial Physics, held in Leningrad, U.S.S.R., 12–19 May 1970.* 1972, VIII + 938 pp.

30. V. Manno and J. Ring (eds.), *Infrared Detection Techniques for Space Research. Proceedings of the 5th ESLAB-ESRIN Symposium, held in Noordwijk, The Netherlands, June 8–11, 1971.* 1972, XII + 344 pp.

31. M. Lecar (ed.), *Gravitational N-Body Problem. Proceedings of IAU Colloquium No. 10, held in Cambridge, England, August 12–15, 1970.* 1972, XI + 441 pp.

32. B.M. McCormac (ed.), *Earth's Magnetospheric Processes. Proceedings of a Symposium Organized by the Summer Advanced Study Institue and Ninth ESRO Summer School, held in Cortina, Italy, August 30–September 10, 1971.* 1972, VIII + 417 pp.

33. Antonin Rükl, *Maps of Lunar Hemispheres.* 1972, V + 24 pp.

34. V. Kourganoff, *Introduction to the Physics of Stellar Interiors.* 1973, XI + 115 pp.

35. B.M. McCormac (ed.), *Physics and Chemistry of Upper Atmospheres. Proceedings of a Symposium Organized by the Summer Advanced Study Institute, held at the University of Orléans, France, July 31–August 11, 1972.* 1973, VIII + 389 pp.

36. J.D. Fernie (ed.), *Variable Stars in Globular Clusters and in Related Systems. Proceedings of the IAU Colloquium No. 21, held at the University of Toronto, Toronto, Canada, August 29–31, 1972.* 1973, IX + 234 pp.

37. R.J.L. Grard (ed.), *Photon and Particle Interaction with Surfaces in Space. Proceedings of the 6th ESLAB Symposium, held at Noordwijk, The Netherlands, 26–29 September, 1972.* 1973, XV + 577 pp.

38. Werner Israel (ed.), *Relativity, Astrophysics and Cosmology. Proceedings of the Summer School, held 14–26 August, 1972, at the BANFF Centre, BANFF, Alberta, Canada.* 1973, IX + 323 pp.

39. B.D. Tapley and V. Szebehely (eds.), *Recent Advances in Dynamical Astronomy. Proceedings of the NATO Advanced Study Institute in Dynamical Astronomy, held in Cortina d'Ampezzo, Italy, August 9–12, 1972.* 1973, XIII + 468 pp.

40. A.G.W. Cameron (ed.), *Cosmochemistry. Proceedings of the Symposium on Cosmochemistry, held at the Smithsonian Astrophysical Observatory, Cambridge, Mass., August 14–16, 1972.* 1973, X + 173 pp.

41. M. Golay, *Introduction to Astronomical Photometry.* 1974, IX + 364 pp.

42. D.E. Page (ed.), *Correlated Interplanetary and Magnetospheric Observations. Proceedings of the 7th ESLAB Symposium, held at Saulgau, W. Germany, 22–25 May, 1973.* 1974, XIV + 662 pp.

43. Riccardo Giacconi and Herbert Gursky (eds.), *X-Ray Astronomy.* 1974, X + 450 pp.

44. B.M. McCormac (ed.), *Magnetospheric Physics. Proceedings of the Advanced Summer Institute, held in Sheffield, U.K., August 1973.* 1974, VII + 399 pp.

45. C.B. Cosmovici (ed.), *Supernovae and Supernova Remnants. Proceedings of the International Conference on Supernovae, held in Lecce, Italy, May 7–11, 1973.* 1974, XVII + 387 pp.

46. A.P. Mitra, *Ionospheric Effects of Solar Flares.* 1974, XI + 294 pp.

48. H. Gursky and R. Ruffini (eds.), *Neutron Stars, Black Holes and Binary X-Ray Sources.* 1975, XII + 441 pp.

49. Z. Švestka and P. Simon (eds.), *Catalog of Solar Particle Events 1955–1969. Prepared under the Auspices of Working Group 2 of the Inter-Union Commission on Solar-Terrestrial Physics.* 1975, IX + 428 pp.

50. Zdeněk Kopal and Robert W. Carder, *Mapping of the Moon.* 1974, VIII + 237 pp.

51. B.M. McCormac (ed.), *Atmospheres of Earth and the Planets. Proceedings of the Summer Advanced Study Institute, held at the University of Liège, Belgium, July 29–August 8, 1974.* 1975, VII + 454 pp.

52. V. Formisano (ed.), *The Magnetospheres of the Earth and Jupiter. Proceedings of the Neil Brice Memorial Symposium, held in Frascati, May 28–June 1, 1974.* 1975, XI + 485 pp.

53. R. Grant Athay, *The Solar Chromosphere and Corona: Quiet Sun.* 1976, XI + 504 pp.

54. C. de Jager and H. Nieuwenhuijzen (eds.), *Image Processing Techniques in Astronomy. Proceedings of a Conference, held in Utrecht on March 25–27, 1975.* 1975, XI + 418 pp.
55. N.C. Wickramasinghe and D.J. Morgan (eds.), *Solid State Astrophysics. Proceedings of a Symposium, held at the University College, Cardiff, Wales, 9–12 July 1974.* 1976, XII + 314 pp.
57. K. Knott and B. Battrick (eds.), *The Scientific Satellite Programme during the International Magnetospheric Study. Proceedings of the 10th ESLAB Symposium, held at Vienna, Austria, 10–13 June 1975.* 1976, XV + 464 pp.